PROCEEDINGS OF
SYMPOSIA IN APPLIED MATHEMATICS
VOLUME I

NON-LINEAR PROBLEMS IN MECHANICS OF CONTINUA

PUBLISHED BY THE
AMERICAN MATHEMATICAL SOCIETY
531 WEST 116th STREET, NEW YORK CITY
1949

PROCEEDINGS OF THE FIRST SYMPOSIUM IN APPLIED MATHEMATICS OF THE AMERICAN MATHEMATICAL SOCIETY

Held at Brown University
August 2–4, 1947

COSPONSORED BY

THE AMERICAN INSTITUTE OF PHYSICS
THE AMERICAN SOCIETY OF MECHANICAL ENGINEERS
THE INSTITUTE OF THE AERONAUTICAL SCIENCES

EDITORIAL COMMITTEE

Eric Reissner
William Prager
J. J. Stoker

TABLE OF CONTENTS

PAGE

Introduction v

Editorial Note vii

HYDRODYNAMICS

Non-linear problems in the theory of fluid motion with free boundaries. By ALEXANDER WEINSTEIN 1

Operator methods in the theory of compressible fluids. By STEFAN BERGMAN 19

An existence theorem in two-dimensional gas dynamics. By LIPMAN BERS 41

Recent developments in free boundary theory. By GARRETT BIRKHOFF 47

Theory of the propagation of shock waves from cylindrical charges of explosive. By STUART R. BRINKLEY, JR. and JOHN G. KIRKWOOD 48

The method of characteristics in the three-dimensional stationary supersonic flow of a compressible gas. By N. COBURN and C. L. DOLPH 55

The numerical solution of the turbulence problem. By HOWARD W. EMMONS 67

On the stability of transonic flows. By Y. H. KUO 72

Stability of the laminar boundary layer in a compressible fluid. By LESTER LEES 74

Remarks on the spectrum of turbulence. By C. C. LIN 81

Two-dimensional compressible flows. By I. OPATOWSKI 87

Polygonal approximation method in the hodograph plane. By H. PORITSKY 94

The boundary layer of yawed cylinders. By W. R. SEARS 117

On shock-wave phenomena: Interaction of shock waves in gases. By H. POLACHEK and R. J. SEEGER 119

The breaking of waves in shallow water. By J. J. STOKER 145

On Hamilton's principle for perfect compressible fluids. By A. H. TAUB 148

ELASTICITY AND PLASTICITY

The foundations of the theory of elasticity. By F. D. MURNAGHAN 158

On dynamic structural stability. By G. F. CARRIER 175

Stress-strain relations for strain hardening materials: Discussion and proposed experiments. By D. C. DRUCKER 181

The edge effect in bending and buckling with large deflections. By K. O. FRIEDRICHS 188

Numerical methods in the solution of problems of non-linear elasticity. By WILFRED KAPLAN 194

Large deflection theory for rectangular plates. By SAMUEL LEVY 197

Discontinuous solutions in the theory of plasticity. By WILLIAM PRAGER 211

On finite deflections of circular plates. By ERIC REISSNER 213

TABLE OF CONTENTS

Introduction [illegible]
Editorial Note [illegible]

HYDRODYNAMICS

Non-linear problems in the theory of fluid motion with free boundaries. By [illegible] Weinstein [illegible]
Operator methods in the theory of compressible fluids. By [illegible] [illegible]
[illegible] flow in two-dimensional gas dynamics. By [illegible] [illegible]
Recent developments in free boundary theory. By [illegible] [illegible]
Theory of the propagation of shock waves from cylindrical charges of explosive. By [illegible] R. Brinkley, Jr. and John G. Kirkwood [illegible]
The method of characteristics in the three-dimensional stationary supersonic flow of a compressible gas. By [illegible] and [illegible] [illegible]
The numerical solution of the [illegible] problem. By [illegible] [illegible]
On the stability of [illegible] flows. By C. C. Lin [illegible]
Stability of the laminar boundary layer in a compressible fluid. By [illegible] Lees [illegible]
[illegible] on the [illegible] [illegible]. By [illegible] [illegible]
Two-dimensional compressible flows. By [illegible] [illegible]
[illegible] approximation [illegible] in the hodograph plane. By [illegible] [illegible]
The boundary layer of [illegible] fluids. By [illegible] [illegible]
[illegible]
[illegible] waves in [illegible] water. By [illegible] [illegible]
[illegible]

[illegible]

[illegible]
[illegible]
[illegible]
[illegible]
[illegible]
[illegible]

INTRODUCTION

In 1946 a group of members of the American Mathematical Society drew attention to the fact that applied mathematics played a smaller part in the activities of the Society than the importance of the subject appeared to warrant. As a result, a Special Committee on Applied Mathematics was appointed to investigate and report. On the basis of the report of this Special Committee, in December 1946 a Committee on Applied Mathematics was appointed, consisting of Professor Richard Courant, Professor G. C. Evans, Professor John von Neumann, Professor William Prager, Dr. Warren Weaver, and the undersigned. To this committee was assigned the general function of encouraging activity in the field of applied mathematics, and the particular duty of organizing an annual Symposium in Applied Mathematics. It was hoped that at such symposia applied mathematicians, primarily affiliated with the American Mathematical Society, would seek the cooperation of physicists, engineers, and others, whose interests were mathematical although they were primarily affiliated to other organizations.

Plans were already being made by Brown University to hold a Symposium in the summer of 1947, and the Committee was happy to recommend the acceptance of the invitation of Brown University to hold the First Symposium in Applied Mathematics of the American Mathematical Society in Providence, using as a basis the Symposium already planned. Most of the duties of organization fell on Professor Prager, and were admirably carried out by him.

It is believed that the Symposia on Applied Mathematics will play an important role in bringing professional mathematicians into contact with mathematical workers in other fields, both purely academic and industrial. To further that end, it is desirable that, while each symposium should focus attention on a fairly restricted field of applied mathematics, the whole sequence should display considerable variety in choice of subject. In view of the fact that the sequence, as planned, is infinite, we may hope to see every phase of applied mathematics adequately covered in the course of time.

JOHN L. SYNGE, *Chairman*
Committee on Applied Mathematics

January 13, 1948

EDITORIAL NOTE

In this volume there are collected the papers which were presented during the First Symposium on Applied Mathematics of the American Mathematical Society. The Symposium was held at Brown University from August 2nd to August 4th, 1947. The subject of the Symposium was *Non-linear problems in mechanics of continua.*

The contents of this volume have been subdivided into two groups, one of them being concerned with the field of Hydro- and Aerodynamics and the other including results in the field of Elasticity and Plasticity.

The decision to publish in one volume the Proceedings of the First Symposium on Applied Mathematics was reached at a time very nearly coinciding with the date of the Symposium. Consequently a few of the papers had already been accepted for publication in the periodical literature. For these papers, published in full elsewhere, the authors have supplied rather comprehensive abstracts.

The Editorial Committee, consisting of William Prager, J. J. Stoker and the undersigned, has been happy to carry out the task of assembling these Proceedings of the First Symposium on Applied Mathematics. We think that this publication expresses well the growing interest in Applied Mathematics in this country in general, and in the American Mathematical Society in particular.

E. Reissner, *Chairman Editorial Committee*
Applied Mathematics Symposium Proceedings

NON-LINEAR PROBLEMS IN THE THEORY OF FLUID MOTION WITH FREE BOUNDARIES

BY

ALEXANDER WEINSTEIN

1. **Introduction.** The outstanding feature of many famous hydrodynamical problems is the somewhat paradoxical fact that the boundary of the flow, on which certain conditions have to be satisfied, is itself not given. There is a great variety of problems with free boundaries, ranging from rotating masses of fluids, investigated already in Newton's time, to shock fronts still to be determined by electronic computers. All these problems are essentially non-linear. Some of them, like those concerning water waves, are problems in the small: one considers only motions which differ but slightly from the state of rest. Other problems however are genuine problems in the large. Among these latter hardly any problem has attracted more attention during the last eighty years than Helmholtz's theory of plane wakes and jets. Some of the mathematical methods developed especially for these problems have already found a wide range of application in other domains of mathematics. Originally the theory was confined to the presentation of numerous examples of such discontinuous flows. The classical work of Helmholtz [1], Kirchhoff [2], Levi-Civita [3], Villat [5] and their followers has been repreatedly presented since 1920–21 in various textbooks on Hydrodynamics. In the present report we confine our attention to some more recent aspects of the theory developed after 1920. Our aim is to summarize present knowledge of the existence and the uniqueness of discontinuous flows with prescribed walls. These questions are still sometimes incorrectly formulated or inadequately treated. Problems of this kind should be of interest not for the mathematician alone. In fact it is still not generally realized in engineering circles that the drag associated with a given body in the theory of cavitation is not necessarily determined by the body alone.

The problem of the wake for plane irrotational motion of an incompressible perfect fluid is the following: Find the flow and the free streamlines of a wake produced by a given obstacle immersed in a parallel flow. The obstacle has either the shape of an open arc or of a closed profile (Fig. 1 and Fig. 9). In the latter case, which is of considerable interest in applications, the separation points of the free boundaries are not prescribed. Since the pressure is constant in and on the boundary of the wake, we see by Bernoulli's equation that the magnitude but not necessarily the direction of the velocity is constant along the unknown free boundaries. For wakes extending to infinity behind the obstacle this constant can be taken equal to one, which is also the speed of the fluid at infinity. The question of the existence of finite wakes seems still to be controversial. In our opinion, the existence of finite wakes is not contradictory to classical hydrodynamics, provided that the pressure in the cavity (as could occur in a cavity filled with air) is higher than the pressure of the adjacent fluid in motion. In this case the constant magnitude of the velocity on the closed free boundary does not necessarily coincide with the speed of the fluid at infinity. A related problem is that of a jet escaping from a given channel. Here again the magnitude of the velocity must be constant of the free boundaries.

There are at present two different methods of approach to the problems of discontinuous motion. The first and older one can be called a *method of continuity*. It can be applied with equal success to jets and to wakes. The second method is a *variational method in the domain of conformal representation*. Its

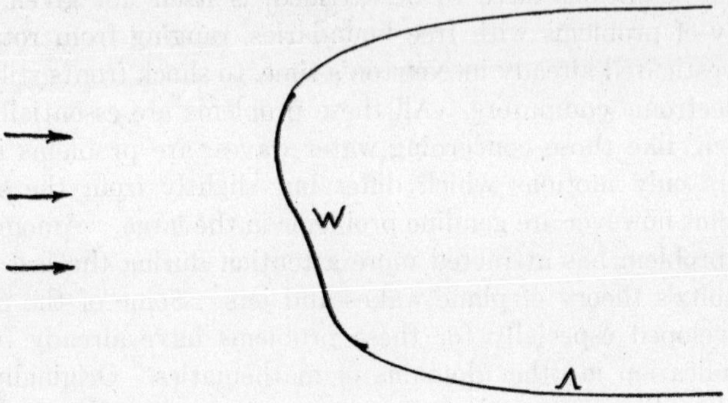

FIG. 1

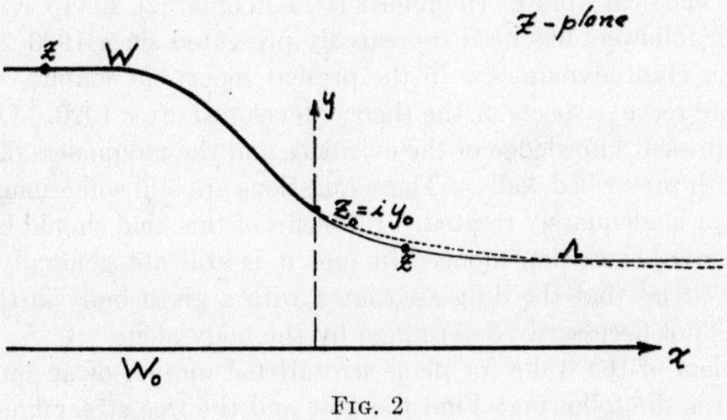

FIG. 2

application seems to be confined to the problem of wakes. The first procedure makes use of the classical semi-inverse method of Levi-Civita and Villat.

2. The inverse problem. Let us consider for the sake of definiteness a symmetric jet in the physical $z = x + iy$ plane. The axis of symmetry can be replaced by a rigid wall and our attention may be confined to the upper part of the jet (Fig. 2). In the following all quantities refer to this part of the flow. Let us put the flux equal to $\pi/2$. Then the speed upstream is known, but *the speed on the free boundary*, which is also the speed downstream at infinity, *is not known*, a fact which has been sometimes overlooked. This constant speed will be denoted by $1/\mu$. The thickness of the jet downstream is then equal to $\mu\pi/2$. Assuming the existence of the flow, let us consider its complex potential $f(z) =$

$\phi + i\psi$. The domain of the jet is mapped conformally in a one-to-one way on a parallel strip S in the f-plane in the manner indicated in Fig. 3. The strip S has the width $\pi/2$ and is bounded by the lines $\psi = 0$ and $\psi = \pi/2$. Corresponding lines are denoted by the same letter and will be called walls and free lines, even if located in the f-plane. Levi-Civita and Villat gave a general solution for the inverse problem, namely a general formula for the inverse function $z(f)$ which is defined in the strip. All functions defined in the physical z-plane can be considered also as functions of f in S. The complex velocity

(1) $$w = df/dz = u - iv$$

considered as a function of f has a constant absolute value on the line Λ which corresponds to the free boundary. Levi-Civita puts

(2) $$dz/df = 1/w = \mu e^{i\omega} = \mu e^{-\tau + i\theta} \qquad (\omega = \theta + i\tau).$$

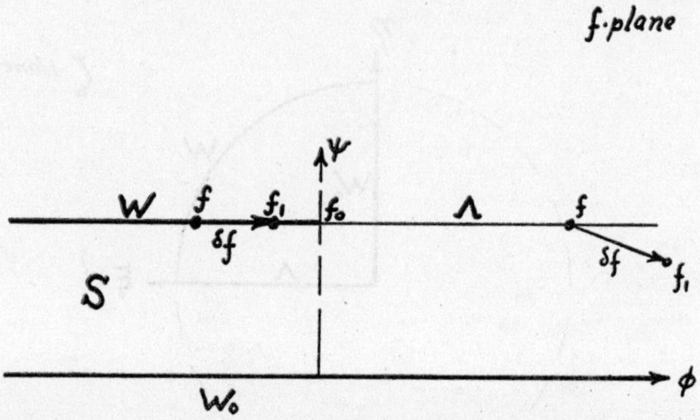

FIG. 3

In this formula, θ denotes the angle which the volocity makes with the x-axis. In the f-plane, θ is a harmonic function of ϕ and ψ, which vanishes on the straight wall W_0. Its conjugate τ vanishes on Λ which implies the condition $d\theta/d\psi = 0$. The values of θ, considered as a function of ϕ and ψ, are not given on the wall W in the f-plane as long as $f(z)$ has not been actually determined. To any arbitrary choice of the function $\theta(\phi, \pi/2)$ on W corresponds a harmonic function $\theta(\phi, \psi)$ vanishing on W_0 and satisfying the condition $d\theta/d\psi = 0$ on Λ. An explicit representation of θ by a Poisson integral can be obtained by using in place of f a new independent variable $\zeta = \xi + i\eta = \rho e^{i\sigma}$ introduced by Levi-Civita and defined by the formulas

(3) $$e^f = \frac{1}{2} i \left(\zeta + \frac{1}{\zeta} \right),$$

(4) $$df = \frac{\zeta^2 - 1}{\zeta^2 + 1} \frac{d\zeta}{\zeta};$$

The one-to-one correspondence between the f and ζ plane is indicated in Fig. 4.

Again the same notations will be used in both planes. By Schwarz' reflection principle the harmonic function θ can be defined in the entire circle $|\zeta| \leqq 1$. For $|\zeta| = 1$, θ satisfies the conditions

$$(5) \qquad \theta(\pi - \sigma) = -\theta(\sigma),$$

$$(6) \qquad \theta(2\pi - \sigma) = \theta(\sigma).$$

Taking arbitrary values for $\theta(\sigma)$ on the wall W in the ζ-plane, we obtain the following formula for $\omega = \theta + i\tau$ (Poisson-Schwarz integral, first used in this connection by Villat):

$$(7) \qquad \omega(\zeta) = \frac{1}{\pi}\int_0^{\pi} \theta(\sigma)\frac{1 - \zeta^2}{1 - 2\zeta\cos\sigma + \zeta^2}\,d\sigma.$$

In this way we see that the general expression for w (considered as a function either of ζ or of f) contains an arbitrary function $\theta(\sigma)$ and an arbitrary positive

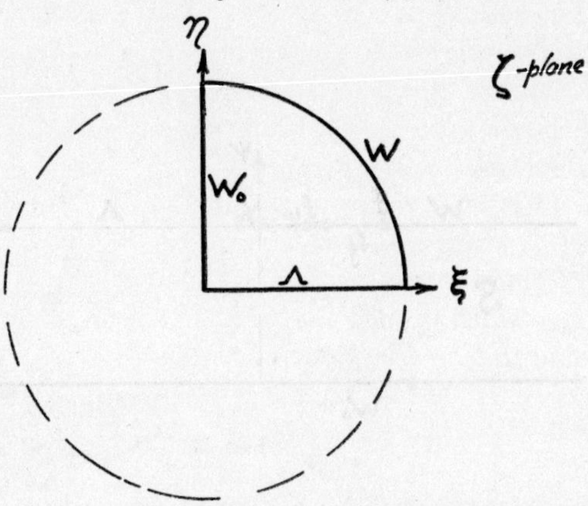

FIG. 4

constant μ. Once $w = \mu^{-1}e^{-i\omega}$ is known as a function of f or of ζ, the function $z(f)$, or, if we wish, $z(\zeta)$, is obtained by a quadrature:

$$(8) \qquad z - iy_0 = \int_{f_0}^{f}\frac{df}{w(f)} = \mu\int_1^{\zeta} e^{i\omega(\zeta)}\frac{\zeta^2 - 1}{\zeta^2 + 1}\frac{d\zeta}{\zeta},$$

where

$$(9) \qquad y_0 = \mathrm{Im}\left[\int_0^{f_0}\frac{df}{w}\right] \qquad (f_0 = i\pi/2)$$

is the ordinate of the separation point of the free line. Before discussing these formulas let us make a remark which will be used later.

Theorem of Boggio. From (7) follows that the harmonic function θ is completely determined by its values on the wall W. In particular, its values on the free boundary lie between the maximum and minimum of its values on W. In view of (6), a level line θ = const. intersects the real axis $\eta = 0$ (that is, the free boundary) in one point only. It follows (Theorem of Boggio) that the free boundary of a jet corresponding to a concave channel (Fig. 2) is necessarily convex with respect to the flow. This proof was communicated to the author by H. Lewy.

The famous general formula (8) of Levi-Civita leads to numerous results obtained by an appropriate choice of the arbitrary function θ along the wall W in the f or ζ plane. However (8) does not yield directly a jet with a prescribed wall in the physical z-plane. This problem will be discussed in the next paragraphs.

3. The functional equation. For a given wall W in the z-plane the angle θ is a given function $\theta = \theta(l)$ of the arc length l measured from the orifice y_0. However the functions $l(\sigma)$ and $\theta(\sigma) = \theta(l(\sigma))$, which appear in (7) and (8), is not known. Our problem is to make the correct choice of the constant μ and of the function $l(\sigma)$. Any arbitrary choice of μ and of the increasing function $l(\sigma)$ will yield a channel which, for increasing l, makes the same succession of angles θ with the x-axis as the given wall, though perhaps not at the same rate. By a proper choice of the positive constant μ we can make the orifices agree, and in the hands of Levi-Civita, Cisotti and Villat this method has proved to be very useful for approximate solutions.

Let us take an arbitrary increasing function $l(\sigma)$ and the corresponding function $\theta(\sigma) = \theta(l(\sigma))$ which we use to define $\omega = \theta + i\tau$ by means of (7). Thus the strip S in the f-plane is mapped into a region in the physical z-plane and gives rise to a channel which is such that the arc length along it, which we shall call $L(\sigma)$, is not equal to $l(\sigma)$. In fact we have by (8)

$$dz = \mu e^{i\omega(\zeta)} \frac{\zeta^2 - 1}{\zeta^2 + 1} \frac{d\zeta}{\zeta}. \tag{10}$$

Along the boundary of the channel $|dz| = dL$. Taking the absolute value of both sides of (10) we obtain

$$\frac{dL}{d\sigma} = \mu e^{-\tau(\sigma)} \tan \sigma. \tag{11}$$

The orifice Y_0 will be given by the right side of (9). Our choice of μ and $l(\sigma)$ is correct only if $L(\sigma) = l(\sigma)$ and $Y_0 = y_0$. Thus we must satisfy the equations

$$\frac{dl}{d\sigma} = \mu e^{-\tau(\sigma)} \tan \sigma, \tag{12}$$

$$y_0 = \operatorname{Im}\left[\mu \int_0^{f_0} e^{i\omega(f)}\, df\right]. \tag{13}$$

Since $l(\sigma)$ occurs in $\tau(\sigma)$ by (7), we have here a very complicated integro-differential equation. For given $\theta(l)$, $l(\sigma)$ and y_0 the function $L(\sigma)$ is determined by (11), in which μ is replaced by its value from (13). Thus we may write

$$L(\sigma) = V[l(\sigma), \theta(l), y_0]. \tag{14}$$

Since $\theta(l)$ and y_0 are fixed for a given channel, the formula (14) is a functional transformation of $l(\sigma)$ into $L(\sigma)$. We shall prove the existence of a flow with prescribed walls if we can find $l(\sigma)$ such that

(15) $$l(\sigma) = V[l(\sigma), \theta(l), y_0].$$

In other words the solution $l(\sigma)$ is a fixed point of the transformation V.

The method of continuity, which has been successfully applied to the problems of jets and wakes, is based on a process of deformation. A wall corresponding to an arbitrary choice of $l(\sigma)$ is continuously deformed into a given shape. The aim of the method of continuity is to show that this process can be followed by a corresponding alteration of the function $l(\sigma)$. The solution of this problem has required several significant changes in the standard procedure of the classical method of continuity.

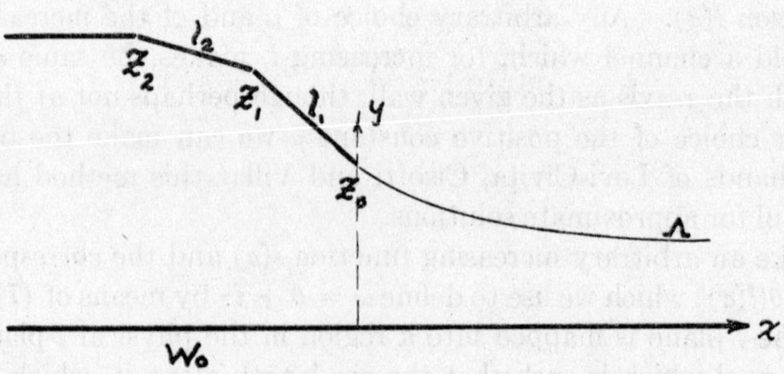

FIG. 5

4. **Polygonal channels.** Let us consider in some detail the case of a polygonal wall, which was the first in chronological order of the problems considered in this theory. Let (Fig. 5)

(16) $$z_0 = iy_0, \qquad z_1 = x_1 + iy_1, \cdots, \qquad z_n = x_n + iy_n, \qquad z_{n+1} = -\infty + iy_n$$

be the vertices of the given polygonal wall of a channel which is concave with respect to the fluid in motion. Let l_k be the length of the segment $z_k z_{k-1}$ ($k = 1, 2, \cdots, n$), and let θ_k be its direction where $\theta_1 \leqq \theta_2 \leqq \cdots \leqq \theta_n \leqq \theta_{n+1} = 0$. Let us put

(17) $$\pi\beta_k = \theta_{k+1} - \theta_k$$

and call

(18) $$\pi \sum_{k=1}^{n} \beta_k = -\theta_1$$

the *total curvature* of the wall. A polygonal channel can be represented by a point P with the coordinates

(19) $$y_0, l_1, l_2, \cdots, l_n, \beta_1, \beta_2, \cdots, \beta_n$$

in a $(2n + 1)$-dimensional space. The region M occupied by the points P is described by the inequalities

$$y_0 > 0, \qquad l_k > 0, \qquad 0 \leqq \beta_k < 1. \tag{20}$$

It follows from (7) that any assumed flow with a polygonal channel is given by taking an arbitrary positive constant μ and by putting

$$\omega(\zeta) = i \sum_{k=1}^{2n} \beta_k \log i \frac{\zeta - \zeta_k}{1 - \zeta\zeta_k} \qquad (\zeta_k = \xi_k + i\eta_k = e^{i\sigma_k}), \tag{21}$$

where the points $\zeta_1, \zeta_2, \cdots, \zeta_n$ $(0 < \sigma_1 < \sigma_2 \cdots < \sigma_n < \pi/2)$ correspond to the vertices $z_1, \cdots, z_n$ of the wall and where $\sigma_{2n-k+1} = \pi - \sigma_k$ and $\theta_{2n+2-k} = -\theta_k$ for $k = 1, 2, \cdots, n$. Inserting (21) into (8) we obtain the *formula of Cisotti*

$$z - iy_0 = \mu e^{i\theta_1} \int_1^\zeta \prod_{k=1}^{2n} \left(\frac{1 - \zeta\zeta_k}{\zeta - \zeta_k} \right)^{\beta_k} \frac{\zeta^2 - 1}{\zeta^2 + 1} \frac{d\zeta}{\zeta} \tag{22}$$

which contains $2n + 1$ parameters

$$\mu, \sigma_1, \sigma_2, \cdots, \sigma_n, \beta_1, \beta_2, \cdots, \beta_n. \tag{23}$$

A channel given by a Cisotti's formula can be therefore also represented by a point P_C in a space of $2n + 1$ dimensions. The region M_C occupied by Cisotti's points P_C is limited by the inequalities

$$\mu > 0, \qquad 0 < \sigma_1 < \sigma_2 < \cdots < \sigma_n < \pi/2, \qquad 0 \leqq \beta_k < 1. \tag{24}$$

To each P_C corresponds a single point P given by the formulas

$$y_0 = \mu\pi \left(1 - \frac{\mu\pi}{y_0}\right) + \mu \sum_{k=1}^{n} \sin \theta_k \int_{\sigma_{k-1}}^{\sigma_k} \prod_{h=1}^{n} \left| \frac{\sin \sigma_h - \sin \sigma}{\sin \sigma_h + \sin \sigma} \right|^{\beta_h} \tan \sigma \, d\sigma,$$

$$l_k = \beta_k = \beta_k \, \mu \int_{\sigma_{k-1}}^{\sigma_k} \prod_{h=1}^{n} \left| \frac{\sin \sigma_h + \sin \sigma}{\sin \sigma_h - \sin \sigma} \right|^{\beta_h} \tan \sigma \, d\sigma \tag{25}$$

$$(k = 1, \cdots, n, \sigma_0 = 0).$$

5. **Existence theorem** (Weinstein [10], Leray and Weinstein [18]). The *existence theorem* for a flow with a given channel states conversely that to each P corresponds at least one P_C, or in other words that the $2n + 1$ non-linear equations (25) possess at least one solution for any prescribed values of the left-hand sides. The introduction of the quantities $\beta_1, \beta_2, \cdots, \beta_n$ as additional parameters, at first glance apparently superfluous, will prove very useful as it will allow us to consider walls with variable sides and variable angles. The existence theorem has been proved for all channels of total curvature less than π; the proof may be outlined as follows. Taking $\beta_1 = \beta_2 = \cdots = \beta_n = 0$ and giving arbitrary values to $\mu, \sigma_1, \cdots, \sigma_n$ we obtain a jet escaping from a channel bounded by a straight wall running parallel to the x-axis. This wall is subdivided into segments of lengths $L_1, \cdots, L_n$ corresponding to the arcs $\sigma_{k-1} \sigma_k$ in the ζ-plane. Let us assume that the following two fundamental theorems have been proved:

THEOREM A. *The functional determinant*

$$D = \frac{\partial(y_0, l_1, \cdots, l_n, \beta_1, \cdots, \beta_n)}{\partial(\mu, \sigma_1, \cdots, \sigma_n, \beta_1, \cdots, \beta_n)} \tag{26}$$

never vanishes in the interior of the region M_C.

THEOREM B. *For all points P located in a closed subregion of the region M, the quantities*

$$\mu, \pi/2 - \sigma_n, \qquad \sigma_1, \sigma_{k+1} - \sigma_k \qquad (k = 1, \cdots, n-1) \tag{27}$$

admit positive upper and lower bounds which are continuous functions of

$$y_0, l_1, \cdots, l_n, \beta_1, \cdots, \beta_n.$$

Theorem A is a consequence of the theorem of local uniqueness and will be discussed later. It holds for all channels of total curvature less than π. Theorem B expresses the a priori limitations of all possible solutions of the equations (25) and follows immediately by an elementary estimation of the right-hand side of these equations. From Theorem A follows that, if a channel $(y_0, l_1, \cdots, l_n, \beta_1, \cdots, \beta_n)$ is given by a formula with the parameters $(\mu, \sigma_1, \cdots, \sigma_n, \beta_1, \cdots, \beta_n)$ the same is true for all neighboring channels given by

$$y_0 + \delta y_0, \qquad l_1 + \delta l_1, \cdots, l_n + \delta l_n, \qquad \beta_1 + \delta\beta_1, \cdots, \beta_n + \delta\beta_n. \tag{28}$$

The corresponding variation of the parameters is obtained by solving the $2n + 1$ linear equations

$$\begin{aligned} \delta y_0 &= \frac{\partial y_0}{\partial \mu} \delta\mu + \sum_{j=1}^{n} \frac{\partial y_0}{\partial \sigma_j} \delta\sigma_j + \sum_{h=1}^{n} \frac{\partial y_0}{\partial \beta_h} \delta\beta_h, \\ \delta l_k &= \frac{\partial l_k}{\partial \mu} \delta\mu + \sum_{j=1}^{n} \frac{\partial l_k}{\partial \sigma_j} \delta\sigma_j + \sum_{h=1}^{n} \frac{\partial l_k}{\partial \beta_h} \delta\beta_h, \\ \delta\beta_k &= \delta\beta_k \qquad\qquad (k = 1, 2, \cdots, n). \end{aligned} \tag{29}$$

Theorem B shows that during the process of deformation of the channel the corresponding parameter point P_C remains in a closed domain in the interior of M_C. It is therefore possible to select a sequence of parameter points P_C converging to a point in M_C which corresponds to the given polygonal channel. The corresponding existence theorem for curved walls may then be obtained by a limiting process [13].

6. **The theory of Leray and Leray-Schauder.** The classical continuity method, as outlined above, is based on the validity of Theorems A and B. However, Brouwer's theory of the topological degree of the mapping of spaces of finite dimension M_C and M yields the following remarkable result: The existence theorem for channels with polygonal walls follows from Theorem B alone, and is independent of the question of validity of Theorem A.

Recently Leray and Schauder [19] gave an important extension of the concept of topological degree to certain completely continuous mappings of Banach spaces. The theory of Leray and Schauder has been developed in view of an immediate application to theory of discontinuous motion, a program which was carried out by Leray [20; 21] for the problem of the wake, which differs in details only from the problem of the jet. Leray considers curvilinear obstacles and shows the existence of at least one fixed point of the transformation analogous to (15). The existence is here again proved independently of the question of local uniqueness and is based on limitations a priori as in Theorem B. No attempt shall be made here to give an exposition of this important and profound theory. We note only that the theorems on the existence of fixed points in function spaces have been obtained by a limiting process starting with the corresponding statements in spaces of a finite number of dimensions. The existence theorem for wakes has been established by Leray for all obstacles which are cut by a parallel to the x-axis in not more than one point or which have with the x-axis an entire segment in common.

7. **Local uniqueness** [7; 10; 8; 9; 12; 15]. Let us now turn our attention to the much discussed question of uniqueness: Is a jet (or wake) uniquely determined by the given rigid wall? Differing from linear problems, the question of uniqueness is here more difficult than the question of existence and there are many cases in which the existence but not the uniqueness of the flow has been proved. Let us first consider the following theorem of *uniqueness in the small* for symmetric jets: There is no second jet infinitely close to the given jet with the same walls but with an infinitesimally different shape of the free boundary. In order to prove this proposition let us denote by $f_1(z)$ the complex potential of a jet which is infinitely close to the given one. Let

$$\delta f = f_1(z) - f(z) \tag{30}$$

be the corresponding variation of $f(z)$. The theorem of uniqueness in the small states that if both jets have the same wall, the function δf vanishes identically:

$$\delta f \equiv 0. \tag{31}$$

The variation δf has been defined here as a function of z in the domain of the given jet. Since we have a one-to-one correspondence between z and $f = f(z)$, where f is the complex potential of the given flow, the variation δf can be considered as a function of f (or of ζ) and can be computed by differentiation of (8) for a fixed value of z. In this way we obtain the equation

$$0 = \frac{\delta f}{w(f)} - \int_{f_0}^{f} \frac{\delta w}{w^2}\, df + i\delta y_0$$

or

$$\delta f = w \int_{f_0}^{f} \frac{\delta \log w}{w}\, df - iw\delta y_0, \tag{32}$$

where δw and δy_0 are the variations of w and y_0 with respect to the parameters $\mu, \sigma_1, \cdots, \sigma_n, \beta_1, \cdots, \beta_n$.

The significance of the equation (31) becomes apparent if we consider first the case of a polygonal wall. Let us consider two neighboring polygonal walls having the same angles at corresponding vertices. By using (21), (8) and (4), we have as a special case of (32)

$$\delta f = e^{-i\omega} \int_1^\zeta e^{i\omega} \delta(-i\omega) \frac{\zeta^2 - 1}{\zeta^2 + 1} \frac{d\zeta}{\zeta} - \frac{\delta\mu}{\mu} \int_1^\zeta e^{i\omega} \frac{\zeta^2 - 1}{\zeta^2 + 1} \frac{d\zeta}{\zeta} - \frac{i\delta y_0}{\mu}, \tag{33}$$

where

$$\delta(-i\omega) = \sum_{k=1}^{2n} i\beta_k \frac{(\zeta^2 - 1)\zeta_k}{(\zeta - \zeta_k)(1 - \zeta\zeta_k)} \delta\sigma_k. \tag{34}$$

The variations $\delta y_0, \delta l_1, \cdots, \delta l_n$ of the orifice and the lengths of the sides of the given wall are connected with the variations of the parameters $\delta\mu, \delta\sigma_1, \cdots, \delta\sigma_n$ by the formulas

$$\begin{aligned} \delta y_0 &= \frac{\partial y_0}{\partial \mu} \delta\mu + \sum_{j=1}^{n} \frac{\partial y_0}{\partial \sigma_j} \delta\sigma_j, \\ \delta l_k &= \frac{\partial l_k}{\partial \mu} \delta\mu + \sum_{j=1}^{n} \frac{\partial l_k}{\partial \sigma_j} \delta\sigma_j \end{aligned} \tag{35}$$

which are obtained by putting $\delta\beta_1 = \cdots = \delta\beta_n = 0$ in (29). Let us assume now that the equation $\delta f \equiv 0$ is a consequence of the equations

$$\delta y_0 = 0, \quad \delta l_1 = 0, \cdots, \quad \delta l_n = 0,$$

which express the fact that the second assumed jet has the same walls as the given one. From (33), (21) and (34) follows immediately that

$$\delta\mu = 0, \quad \delta\sigma_1 = 0, \cdots, \quad \delta\sigma_n = 0. \tag{36}$$

This result means that the system of homogeneous equations obtained by putting in (35) the left-hand sides equal to zero admits only the trivial solution (36). The equation (31) yields therefore the following result. *The determinant of the system* (35), *which is obviously equal to the functional determinant* (26) *in Theorem* A, *is always different from zero.*

In the case of curvilinear walls Leray has shown the importance of the equation (31) by establishing the connection existing between the variation δf considered by Weinstein and the differential, in the sense of Fréchet, of the functional transformation (14).

As in the case of polygonal walls, the vanishing of δf yields the result that a fixed point of the transformation (14) is a locally isolated solution of (15).

8. **Boundary value problem** (Weinstein [7; 10]). The proof of the fundamental equation (31) which plays a central role in this theory has been established by studying the boundary conditions satisfied by the variation δf. Since, by assumption, both jets have the same wall, the points $f = f(z)$ and $f_1 = f_1(z)$ corre-

sponding to the same point z on the wall W (or W_0) in the physical z-plane both lie on the line W (or W_0) in the f-plane so that the variation δf is real (Fig. 3). We have therefore the boundary condition

$$\delta\psi = 0 \tag{37}$$

on W and W_0 in the f-plane. This reasoning would obviously fail on the given free boundary because a point z on the first free boundary Λ does not necessarily lie on the second free boundary (Fig. 2). Fortunately we have there a natural boundary condition which holds no matter whether we have the same rigid wall or not for both jets. In fact from (32) follows immediately by differentiation the fundamental identity

$$\frac{d(\delta f)}{df} - \frac{d \log w}{df}\,\delta f = \delta \log w, \tag{38}$$

the real part of which on the free boundary in the f-plane yields the following natural boundary condition:

$$\frac{d(\delta\psi)}{dn} - \frac{d\theta}{d\phi}\,\delta\psi = -\delta \log \mu \qquad (dn = d\psi) \tag{39}$$

where $(1\mu)(d\theta/d\phi)$ is the curvature K of Λ. Moreover, for an unvaried separation point, the imaginary part of the same identity yields the supplementary condition

$$\frac{d(\delta\psi)}{d\phi} = 0 \tag{40}$$

for $f = f_0$. We note that the unknown constant $\delta\mu$ appears explicitly in (39). The uniqueness in the small is thus reduced to proving the following

THEOREM I (LOCAL UNIQUENESS). *The harmonic function* $\delta\psi$ *satisfying the boundary conditions* (37), (39) and (40) *is identically zero.*

The proof of I has been obtained by considering first the special case in which $\delta\mu$ is equal to zero. This case is that of neighboring jets having not only the same walls but also the same width at infinity downstream. This assumption is an artificial restriction but, as will be indicated, the complete proof of the fundamental Theorem I may be shown to follow from this simpler and apparently special case.

Putting $\delta\mu = 0$ and denoting in this case the corresponding harmonic function $\delta\psi$ by β, we formulate the following special case of I:

THEOREM II (LOCAL UNIQUENESS IN THE RESTRICTED SENSE). *The harmonic function* $\beta(\phi, \psi)$ *satisfying the boundary conditions*

$$\beta = 0 \text{ on } W \text{ and } W_0 \tag{41}$$

(42) $$\frac{d\beta}{dn} = \frac{d\theta}{d\phi}\beta \text{ on } \Lambda$$

vanishes identically in the strip S.

A classical application of Green's formula, due to Fourier, leads to the equation

(43) $$\int\int_S \left\{\left(\frac{\partial\beta}{\partial\phi}\right)^2 + \left(\frac{\partial\beta}{\partial\psi}\right)^2\right\} d\phi\, d\psi - \int_\Lambda \frac{d\theta}{d\phi}\beta^2\, ds = 0.$$

The proof of II would be trivial if the function $d\theta/d\phi$ were shown to be always negative, but unfortunately in our problem this usually is not the case. In fact, in the important case of a channel with walls concave with respect to the fluid, $d\theta/d\phi$ is always positive (see Theorem of Boggio); in this case the left-hand side in (43) is the difference and not the sum of two positive numbers. The proof of Theorem II presents therefore a major difficulty. Proofs valid for more and more extensive classes of given walls have been established successively by Weinstein [7], Hamel [8; 9], Weyl [12] and Friedrichs [15].

9. **Jacobi's transformation** [15]. The best and latest proof of II is that of Friedrichs, who showed that under certain conditions the left-hand side of (43) can be written as the sum of two positive quantities. This transformation holds provided that there exists a harmonic function U in S which is always different from zero and which satisfies the natural boundary condition

(44) $$\frac{dU}{dn} = \frac{d\theta}{d\phi} U \text{ on } \Lambda.$$

Assuming the existence of U, Friedrichs puts in (43) $\beta = U\eta$ (Jacobi's multiplicative variation). A simple computation shows then that (43) is transformed into the equation

(45) $$\int\int_S U(\eta_\phi^2 + \eta_\psi^2)\, d\phi\, d\psi = 0$$

from which follows immediately the required equation $\eta \equiv 0$ and $\beta \equiv 0$. Let us consider now the question of the existence of the harmonic function U. It has been pointed by C. Jacob that the components of the velocity $u = \mu^{-1} e^\tau \cos\theta$ and $v = \mu^{-1} e^\tau \sin\theta$ satisfy the natural boundary condition (44), a fact easily verified by differentiation. Let us assume the existence of a component $au + bv$ (where a and b are two constants satisfying the equation $a^2 + b^2 = 1$) which never vanishes on the given wall W. Then, by the Theorem of Boggio, this component is everywhere different from zero and we may take for U the function

(46) $$U = au + bv.$$

It follows that Theorem II is valid for a large class of channels, in particular for all concave channels with walls of total curvature less than π.

10. **Reduction of Theorem I to Theorem II.** Let us consider a channel for which the local uniqueness in the restricted sense (Theorem II) has been established. We shall prove that II implies I (Weinstein [10], Leray [21]). Let us introduce in I as new unknown the harmonic function $\beta^* = \delta\psi + (\psi - \pi/2)\delta \log \mu$. We have, by (37), (39) and (40) the following boundary conditions for β^*

$$\beta^* = -(\pi/2)\, \delta \log \mu \text{ on } W_0, \tag{47}$$

$$\beta^* = 0 \text{ on } W, \tag{48}$$

$$\frac{d\beta^*}{dn} = \frac{d\theta}{d\phi}\beta^* \quad \text{on} \quad \Lambda, \tag{49}$$

$$\frac{d\beta^*}{d\phi} = 0, \qquad \frac{d\beta^*}{d\psi} = 0 \quad \text{for } f = f_0. \tag{50}$$

The conditions (48) and (50) show the existence of a level line $\beta^* = 0$ different from W and passing through the separation point f_0. This line must have another point in common either with W or with W_0 or with the free boundary Λ. In the first case we have immediately the result $\beta^* \equiv 0$. In the second case we would have $\delta\mu = 0$ so that β^* would satisfy the conditions of II and therefore would vanish identically. Finally in the third case β^* satisfies the conditions of II in a certain subdomain S^* of S, which again yields the result $\beta^* \equiv 0$. Since $\beta^* = 0$ implies $\delta\mu = 0$ and $\delta\psi \equiv 0$, Theorem I is proved.

11. **Uniqueness in the large** [18]. Let us consider again for the sake of definiteness the class of all polygonal channels for which the existence of a jet and the local uniqueness have been established. The uniqueness in the large or the uniqueness in the ordinary sense is obvious in the trivial case of a straight wall running parallel to the x-axis. The corresponding points in the domains M and M_C will be denoted by O and O_C respectively. Let us assume for a moment the existence of two points $P_C^{(1)}$ and $P_C^{(2)}$ in M_C corresponding to a single point P in M. We shall establish a contradiction by linking P with O by a line in M which lies in the interior of a closed subdomain of M. To this line correspond two lines $P_C^{(1)}O_C$ and $P_C^{(2)}O_C$ in M_C which lie, by Theorem B, in the interior of a closed subdomain of M_C. Let us consider the first point Q_C which these lines have in common. Let Q be the corresponding point on the line PO. Clearly Theorem I does not hold for Q. We have thus a contradiction with the assumptions, and the uniqueness in the large is therefore proved. The uniqueness in the large for curvilinear walls has been established along the same lines by Leray [21] in the theory of wakes. His proof is valid for convex obstacles and for symmetric obstacles whether convex or not. The only essential assumption is again that the obstacle is cut by a parallel to the main stream in not more than one point. It should be noted that the velocity at infinity downstream on the free lines of a wake is always parallel to the x-axis, whether the obstacle is symmetric or not. The situation is, however, different in the theory of jets with nonsymmetric channels, where the direction of the velocity downstream is not given. For this reason the problem of the nonsymmetric jets has not yet been solved.

12. Lavrentieff's method. The questions just discussed belong to the first method of approach to the theory of jets and wakes. More recently, just before the war, Lavrentieff [30] outlined an entirely new method in a formidable and difficult paper. Lavrentieff's method seems to be applicable only to *symmetric wakes* and not to jets. It will suffice to consider the lower part of the wake. The fundamental assumption is that the obstacle is cut by every vertical (not horizontal) in not more than one point, and therefore can be represented by an equation $y = F(x)$, which is an assumption quite different from the one just considered. An outstanding feature of Lavrentieff's work is that he does not use the inverse method of Levi-Civita and Villat. The following sketch is very incomplete and is intended to give only a general idea of Lavrentieff's methods.

Lavrentieff's existence theorems. The existence is proved from the solution of the following variational problem: to find, for a given obstacle, a free boundary Λ such that the total variation of $| df/dz |$ is a minimum along Λ. Lavrentieff shows that this minimum is equal to zero, so that $| df/dz |$ remains constant along Λ as required by the solution of the problem. The proof involves numerous and profound investigations which have great value for the general theory of conformal representation. Let us consider in some detail the question of the *uniqueness in the large*, the uniqueness in the small being here no longer a necessary prerequisite. Assume a second wake with the same obstacle and assume for the sake of the argument that the new free line of the second wake is completely below the given one.

Considering the lower half of the flow, we have then two domains D_1 and D_2 limited by W_0, W, Λ_1 and W_0, W, Λ_2 respectively (see Fig. 6). D_1 and D_2 are mapped by the corresponding potentials $f_1(z)$ and $f_2(z)$ on the same half-plane, $f_1'(z)$ and $f_2'(z)$ being both equal to one for $z = \infty$. Moreover D_2 is completely contained in D_1, but has a portion of the boundary in common. Now Lavrentieff proves the following fundamental lemma: Under the circumstances just enumerated for D_1 and D_2, the modulus $| f_2'(z) |$ is less than the modulus $| f_1'(z) |$ at any point of the boundary which is common to both domains. (The bigger domain has the bigger derivative in modulus.) If we assume however that these two lines are boundaries of the two wakes, we would have $| f_1' | = | f_2' |$ at the separation point, which would contradict the lemma and therefore clash with the assumption of two different wakes. A refinement of this argument can be applied to the general case when the two assumed boundaries intersect.

Lavrentieff considers also closed symmetric convex profiles and shows, among other results, that there is on the front side an arc such that the profile lies entirely in the wake corresponding to this arc. To show this, Lavrentieff considers Kirchhoff's wake for a plate touching the front of the profile (Fig. 7). Lavrentieff takes the plate of such a size that the Kirchhoff wake just touches the profile (obviously somewhere at the front side) without cutting the profile. Let us now consider the wake due to the arc of the profile which lies between the two points of contact, these points being the separation points for this wake.

Let us denote by A the lower separation point. Lavrentieff shows that the free boundary of this wake lies completely outside of the Kirchhoff wake and that therefore it cannot intersect the closed profile (Fig. 7). To give an idea of his proof let us show at least that the profile wake cannot lie completely inside the Kirchhoff wake. In fact, such a configuration would lead again to two domains (Fig. 8) bounded by W_0, K, Λ_K and by W_0, W, Λ respectively, one of

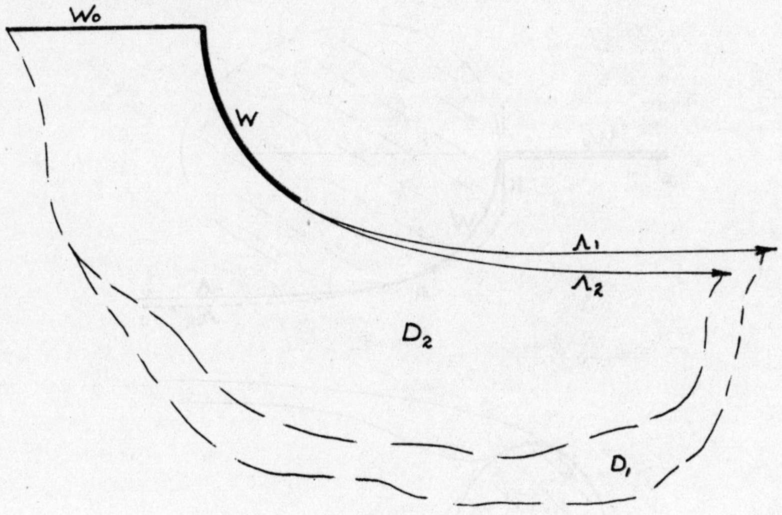

FIG. 6

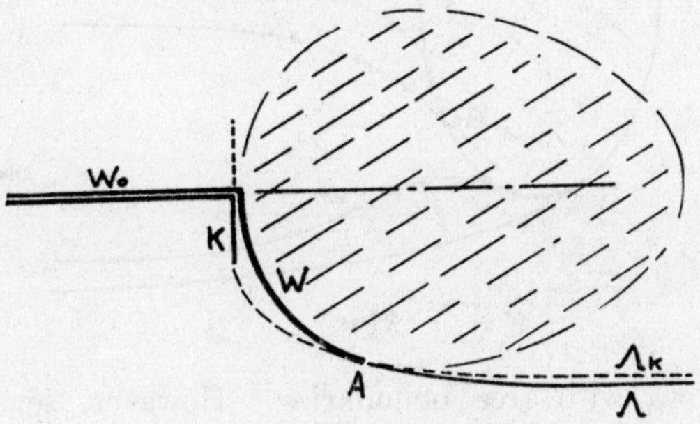

FIG. 7

which is completely contained in the other, both of them being mapped on the same f-half-plane, the derivative being one at infinity. But then, according to the lemma, the speed could not be equal to one on both free boundaries at the point of contact, which proves that the assumed configuration is impossible.

The general result, including finite wakes, is illustrated by Fig. 9. Lavrentieff finds that the constant speed on the free boundary of a finite wake is less than the speed of the flow at infinity. The proofs given by Lavrentieff involve the separate discussion of several possible configurations of the free lines.[1] It should be frankly stated here that the present reviewer did not check in detail the stringency of all Lavrentieff's arguments. Lavrentieff's theory seems to be not applicable to the theory of jets because an important statement, formulated as the second part of his Theorem I, is no longer valid in this case.

13. Concluding Remarks. In the present report, incomplete as it is, an attempt has been made to sketch the great achievements of the last 25 years in

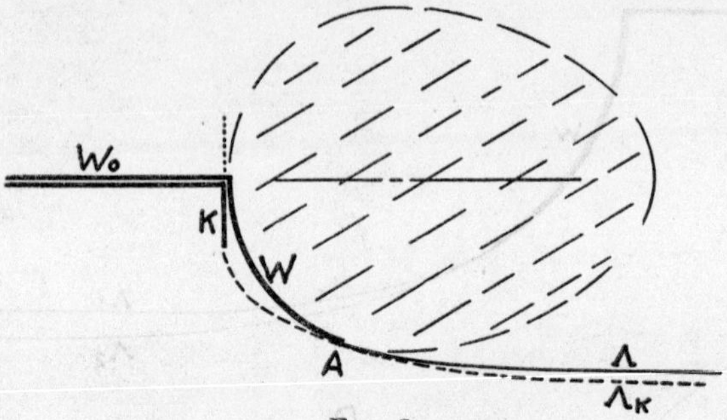

FIG. 8

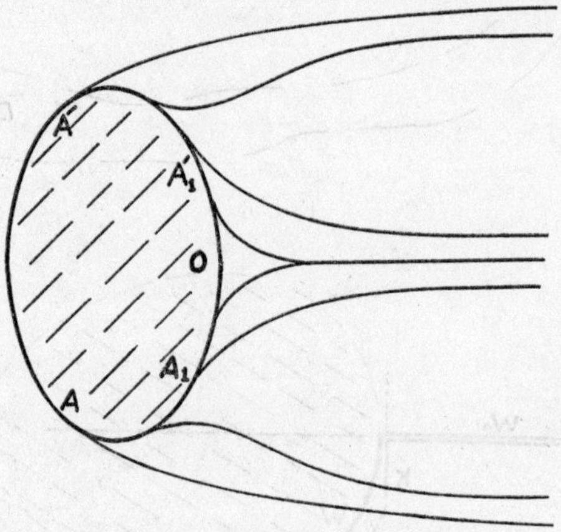

FIG. 9

the theory of flows with free boundaries.[2] However, several outstanding problems remain, inasmuch as the method of continuity and the method of Lavrentieff prove existence and uniqueness for different classes of given walls

[1] An example of an obstacle with different points of separation had been given previously by Bergman [14].

[2] Among recent contributions should be mentioned the work of Kravtchenko [33; 35], who extended the topological method of continuity to more general configurations, and the unpublished work of M. Schiffer (finite wakes). One of Kravtchenko's papers [33] contains an extensive bibliography, and also mentions some authors like Schmieden and Sekorj-Zenkowitch who, while still quoted in modern literature as contributors to the subject, have committed serious errors. See also the comment of Leray [21, p. 153] on Schmieden.

and in many cases the results established by one of these methods cannot be obtained by the other. Brilliantly ingenious as is the method of Lavrentieff, on the whole it seems that the method of continuity furnishes a greater contribution to the general development of the mathematics of non-linear problems.

Bibliography

1. H. Helmholtz, *Ueber discontinuierliche Flüssigkeitsbewegungen*, Monatsberichte der Berliner Akademie, 1868.

2. G. Kirchhoff, *Zur Theorie freier Flüssigkeitsstrahlen*, J. Reine Angew. Math. vol. 70 (1869) p. 289.

3. T. Levi-Civita, *Scie e leggi di resistenza*, Rend. Circ. Mat. Palermo vol. 23 (1907) p. 1.

4. H. Villat, *Sur la résistance des fluides*, Ann. École Norm. vol. 28 (1911) p. 203.

5. ———, *Aperçus théoriques sur la résistance des fluides*, Collection Scientia, Paris, 1920.

6. U. Cisotti, *Idrodinamica Piana*, Milan, 1921.

7. A. Weinstein, *Sur l'unicité des mouvements glissants*, C. R. Acad. Sci. Paris vol. 176 (1923) p. 493; *Ein hydrodynamischer Unitätssatz*, Math. Zeit. vol. 19 (1924) p. 265.

8. G. Hamel, *Ueber einen hydrodynamischen Unitätssatz des Herrn Weinstein*, Résumé des Conférences du Deuxième Congrès International de Mécanique appliquée, Zurich, 1926, p. 76.

9. ———, *Ein hydrodynamischer Unitätssatz*, Proceedings of the Second International Congress of Applied Mechanics, Zurich, 1926, p. 489.

10. A. Weinstein, *Sur les jets liquides a parois données*, Rendiconti d. R. Accademia dei Lincei, 1926, p. 119; *Sur le théorème d'existence des jets fluides*, ibid. 1927, p. 157.

11. E. G. C. Poole, *On the problem of two obstacles in an infinite stream*, Proc. London Math. Soc. vol. 26 (1927) p. 148.

12. H. Weyl, *Strahlbildung, nach der Kontinuitätsmethode behandelt*, Nachr. Ges. der Wiss. Göttingen, 1927, p. 227.

13. A. Weinstein, *Zur Theorie der Flüssigkeitsstrahlen*, Math. Zeit. vol. 31 (1929) p. 424.

14. S. Bergman, *Mehrdeutige Lösungen bei Potentialströmungen mit freien Grenzen*, Zeitschrift für angewandte Mathematik und Mechanik vol. 12 (1932) p. 83.

15. K. Friedrichs, *Ueber ein Minimumproblem mit freiem Rande*, Math. Ann. vol. 109 (1933) p. 64.

16. A. Weinstein, *Sur les sillages provoqués par des arcs circulairés*, Rendiconti d. R. Accademia d. Lincei vol. 17 (1933) p. 83; *Sur les points de detachement des lignes de glissement*, C. R. Acad. Sci. Paris vol. 196 (1933) p. 324.

17. B. Demtchenko, *Problèmes mixtes harmoniques en hydrodynamique des fluides parfaits*, Paris, 1933.

18. J. Leray and A. Weinstein, *Sur un problème de représentation conforme pose par la théorie de Helmholtz*, C. R. Acad. Sci. Paris vol. 198 (1934) p. 430.

19. J. Leray and J. Schauder, *Topologie et équations fonctionnelles*, Ann. École Norm. vol. 51 (1934) p. 45.

20. J. Leray, C. R. Acad. Sci. Paris vol. 199 (1934) p. 1282; vol. 200 (1935) p. 2007.

21. ———, *Les problèmes de représentation conforme de Helmholtz*, Comment. Math. Helv. vol. 8 (1935–1936) pp. 149 and 250.

22. ———, *Sur la validité des solutions du problème de la proue*, Volume du Jubilé de M. M. Brillouin, 1935, p. 246.

23. C. Jacob, *Sur la détermination des fonctions harmoniques par certaines conditions aux limites*, Mathematica vol. 11 (1935) p. 150.

24. A. Weinstein, *Les conditions aux limites introduites par l'hydrodynamique*, L'Enseignement Mathématique vol 35 (1936) p. 107 (contains an extensive bibliography).

25. V. Valcovici, *Sur le sillage derrière un obstacle circulaire*, Comptes Rendus Congrès International des Mathématiciens, vol. 2, Oslo, 1936, p. 250.

26. J. Pérès, *Cours de Mécanique des fluides*, Paris, 1936.

27. R. Courant and D. Hilbert, *Methoden der mathematischen Physik*, vol. 2, Berlin, 1937, p. 175.

28. H. Villat, *Mécanique des fluides*, 2d ed., Paris, 1938.

29. J. Kravtchenko, C. R. Acad. Sci. Paris, vol. 200 (1935) p. 208; vol. 200 (1935) p. 1832; vol. 201 (1936) p. 250; vol. 203 (1936) p. 426; vol. 205 (1937) p. 1203.

30. M. Lavrentieff, C. R. (Doklady) Acad. Sci. URSS. vol. 18 (1938) p. 225.

31. ———, *Sur certaines propriétes des fonctions univalentes et leurs applications à la théorie des sillages*, Rec. Math. (Mat. Sbornik) vol. 46 (1938) pp. 391–458.

32. Th. von Kármán, *The engineer grapples with nonlinear problems*, Bull. Amer. Math. Soc. vol. 46 (1940) pp. 616–683 (contains an extensive bibliography).

33. J. Kravtchenko, *Sur le problème de représentation conforme de Helmholtz; théorie des sillages et des proues*, J. Math. Pures Appl. vol. 20 (1941) pp. 35–303 (contains an extensive bibliography).

34. A. Oudart, *Sur le schéma de Helmholtz-Kirchhoff*, J. Math. Pures Appl. vol. 22 (1943) p. 245; vol. 23 (1944) p. 1.

35. J. Kravtchenko, *Sur l'existence des solutions du problème de représentation conforme de Helmholtz*, Ann. École Norm. vol. 62 (1945) pp. 233–268 and vol. 63 (1946) pp. 161–184.

36. M. Shiffman, *On free boundaries of an ideal fluid*, Communications on Applied Mathematics, New York University, vol. 1 (1948) p. 89.

Naval Ordnance Laboratory,
Silver Spring, Md.
University of Maryland,
College Park, Md.

OPERATOR METHODS IN THE THEORY OF COMPRESSIBLE FLUIDS[1]

BY

STEFAN BERGMAN

1. **Partial differential equations arising in the theory of two-dimensional flows of a compressible fluid.** The fact that the mathematical theory of steady two-dimensional flows of an incompressible fluid is essentially identical with the investigation of certain problems in the theory of functions of one complex variable makes it possible to use the highly developed tools of the latter theory; this is one of the main reasons for the considerable success in the development of this theory. In connection with the important problems arising from applications in aeronautics, there has recently developed a considerable interest in studying two-dimensional *compressible* fluid flows. The success in the incompressible fluid case suggests the generalization of the aforementioned methods to the latter theory. If we wish to use such an approach, certain modifications are needed. In the case of an incompressible fluid we can operate either in the physical plane (that is, consider the stream function ψ as a function of the coordinates x, y of the plane where the flow actually takes place) or in the hodograph plane in which one considers ψ as a function of the velocity vector (hodograph method). In the case of an incompressible fluid, $\psi(x, y)$ as well as $\psi(\lambda^{\blacktriangle}, \theta)$, $\lambda^{\blacktriangle} = \log v$, v being the speed, and θ the angle which the velocity vector makes with some fixed direction, *both* satisfy Laplace's equation.

Since the transition to the hodograph plane distorts many relations, it is as a rule more convenient in the incompressible fluid case to operate directly in the physical plane. In the case of a compressible fluid $\psi(x, y)$ satisfies a complicated *non-linear* equation, while (in the hodograph plane) ψ satisfies a linear one.

In particular, in the case of isentropic flow (with the pressure of the form $p = \sigma\rho^k$, σ and k being constants and p and ρ being pressure and density) it is convenient to consider ψ in the H, θ-plane where

$$\text{H} = \int_{v_1}^{v} v^{-1}\left[1 - \frac{1}{2}a_0^{-2}(k-1)v^2\right]^{1/(k-1)} dv, \tag{1.1}$$

v_1 being the speed corresponding to the Mach number $M = 1$. The compressibility equation then assumes the form:

$$S(\psi) = l(\text{H})(\partial^2\psi/\partial\theta^2) + (\partial^2\psi/\partial\text{H}^2) = 0, \qquad l(\text{H}) = (1 - M^2)/\rho^2 \tag{1.2}$$

[4, (42)]. The function $l(\text{H})$ is positive for $M < 1$ and negative for $M > 1$.

---

[1] Research paper done under Navy Contract NOrd 8555-Task F, at Harvard University. The ideas expressed in this paper represent the personal view of the author and are not necessarily those of the Bureau of Ordnance.

REMARK 1.1. At the point H = 0 (which corresponds to $M = 1$), $l(\mathrm{H})$ has the development

$$
(1.3) \quad \begin{aligned} l(\mathrm{H}) &= \left(\frac{2}{k-1}\right)^{(2-k)/(k-1)} \\ &\cdot \left[(-2\mathrm{H}) - \left(\frac{2k+5}{2k+2}\right)\left(\frac{k+1}{2}\right)^{2k/(k-1)} (-2\mathrm{H})^2 + \cdots\right]. \end{aligned}
$$

Since the theory of non-linear partial differential equations is not sufficiently developed, it is more convenient for many purposes to use the hodograph method in the case of a compressible fluid despite the fact that the transition to the hodograph plane brings, in many instances, considerable distortion of the relations in which we are interested.

It is well known that the compressibility equation in the supersonic case becomes one of hyperbolic type and that the behavior of solutions of the latter is completely different from that of solutions of equations of elliptic type. Therefore, if we wish to have a strong analogy to the incompressible fluid case (which is governed by Laplace's equation), we have to limit ourselves to the subsonic region so that we are led to consider at first the streamfunctions of *subsonic* flows and in the hodograph plane. In the subsonic case, we can reduce (1.2) to the canonical form for equations of elliptic type, introducing instead of H a new variable.

$$
(1.4) \quad \begin{aligned} \lambda &= \frac{1}{2}\lg\frac{1-(1-M^2)^{1/2}}{1+(1-M^2)^{1/2}} + \frac{1}{2h}\lg\frac{1+h(1-M^2)^{1/2}}{1-h(1-M^2)^{1/2}}, \\ h &= \left(\frac{k-1}{k+1}\right)^{1/2} \end{aligned}
$$

[4,§ 7]. (1.2) then becomes

$$
(1.5) \quad \frac{\partial^2\psi}{\partial\lambda^2} + \frac{\partial^2\psi}{\partial\theta^2} + 4N\frac{\partial\psi}{\partial\lambda} = 0, \quad N = -\frac{k+1}{8}\,\frac{M^4}{(1-M^2)^{3/2}}.
$$

REMARK 1.2. Often instead of ψ we consider $\psi^* = \psi/H$, where

$$
(1.6) \quad \begin{aligned} H &= (1-M^2)^{-(1/4)}\,[1 + 2^{-1}(k-1)M^2]^{-1/2(k-1)} \\ &= S_0\,(-2\lambda)^{-1/6}\,[1 + S_1(-2\lambda)^{2/3} + S_2(-2\lambda)^{4/3} + \cdots] \end{aligned}
$$

[9, (4.3)]. ψ^* satisfies a somewhat simpler equation than (1.5) namely

$$
(1.7) \quad \frac{\partial^2\psi^*}{\partial\lambda^2} + \frac{\partial^2\psi^*}{\partial\theta^2} + 4F\psi^* = 0,
$$

where

$$
(1.8) \quad \begin{aligned} 4F &= \frac{(k+1)M^4\,[-(3k-1)M^4 - 4(3-2k)M^2 + 16]}{16(1-M^2)^3} \\ &= \frac{1}{\lambda^2}\,[\alpha_0 + \alpha_1(-\lambda)^{2/3} + \cdots], \qquad \alpha_0 = \frac{5}{36}, \alpha_1 = 0, \cdots \end{aligned}
$$

[4, (70), (71)]. For the potential function we obtain the equation

$$\frac{\partial^2 \phi}{\partial \lambda^2} + \frac{\partial^2 \phi}{\partial \theta^2} - 4N\frac{\partial \phi}{\partial \lambda} = 0. \tag{1.9}$$

REMARK 1.3. It should perhaps be added that in certain instances it is preferable also in the mixed case (that is, in the case when the motion is partially sub-

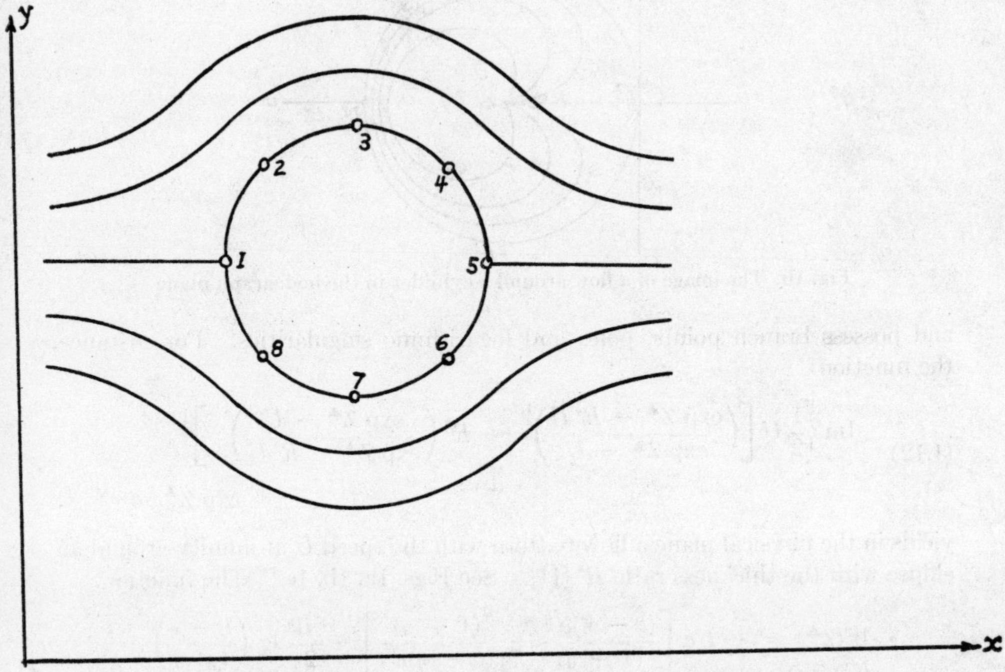

Fig. 1a. The image of a flow around a cylinder in the physical plane

sonic and partially supersonic) to use the variables λ, θ. Since, for $M > 1$, the function λ introduced in (1.4) becomes

$$-\lambda(M) = -i\,[\arctan\,(M^2-1)^{1/2} - h^{-1}\arctan\,(h(M^2-1)^{1/2})] \equiv i\Lambda(M), \tag{1.10}$$

it is useful to extend the function under consideration to *complex* values of the argument λ. Let us introduce $\lambda_1 + i\Lambda$ instead of λ, and then omit the subscript "1" in λ_1. Considering $\psi(\lambda + i\Lambda, \theta)$, we have $\Lambda = 0$ in the subsonic region and $\lambda = 0$ in the supersonic region. In the first case $\psi(\lambda, \theta)$ satisfies equation (1.5) and in the second case the equation

$$-\frac{\partial^2 \psi}{\partial \Lambda^2} + \frac{\partial^2 \psi}{\partial \theta^2} - 4N_1\frac{\partial \psi}{\partial \Lambda} = 0, \quad N_1 = \frac{k+1}{8}\,\frac{M^4}{(M^2-1)^{3/2}}. \tag{1.11}$$

A visualization of this situation is indicated in Fig. 3b, p. 35, where the (λ, Λ)-plane is chosen horizontal, while the vertical direction is chosen as that of the θ axis: The half plane $[\Lambda = 0, \lambda < 0, -\infty < \theta < \infty]$ corresponds to the subsonic region, while $[\lambda = 0, \Lambda > 0, -\infty < \theta < \infty]$ corresponds to the supersonic region. Both half planes are connected along the axis: $\lambda = 0, \Lambda = 0$.

In the case of an incompressible fluid, stream functions $\psi(\lambda^\blacktriangle, \theta)$ which yield in the physical plane flow patterns around closed curves are, in general, multivalued

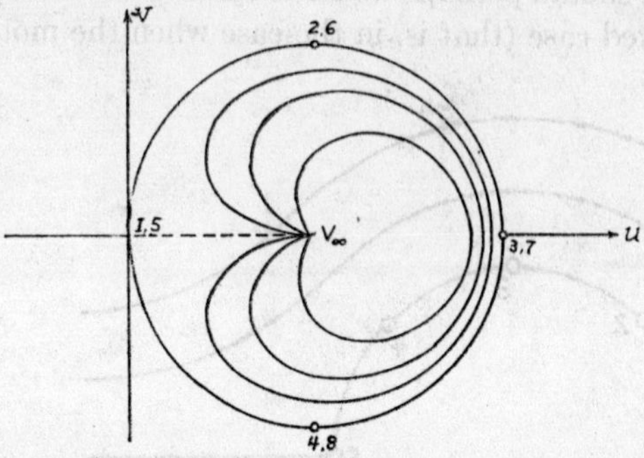

Fig. 1b. The image of a flow around a cylinder in the hodograph plane

and possess branch points, poles and logarithmic singularities. For instance, the function

$$\text{(1.12)} \quad \operatorname{Im}\left\{\frac{1}{2} U \left[\left(\frac{\exp Z^\blacktriangle - R^2 U}{\exp Z^\blacktriangle - U}\right)^{1/2} + R^2 \left(\frac{\exp Z^\blacktriangle - U}{\exp Z^\blacktriangle - R^2 U}\right)^{1/2}\right]\right\},$$

$$\exp Z^\blacktriangle \equiv e^{\lambda^\blacktriangle - i\theta},$$

yields in the physical plane a flow pattern with the speed U at infinity around an ellipse with the thickness ratio R^2 [11]. See Figs. 1a, 1b, 1c.[2] The function

$$W(Z^\blacktriangle) = -Ua\left[\frac{(t-\eta)e^{i\alpha}}{(1+\eta)} + \frac{(1+\eta)}{(t-\eta)e^{i\alpha}}\right] - \frac{i\Gamma}{2\pi} \lg\left(\frac{t-\eta}{1+\eta}\right),$$

where $t = t(Z^\blacktriangle)$ is a solution of

$$\exp(Z^\blacktriangle) = \left(\frac{-t^2}{t^2-1}\right)\left[\frac{e^{i\alpha}U}{1+\eta} + \frac{U(1+\eta)}{(t-\eta)^2 e^{i\alpha}} + \frac{i\Gamma}{2\pi a}\frac{1}{(t-\eta)}\right]$$

and U, α, η, Γ are constants, yields a flow pattern around a Joukowski profile [6, pp. 55–57]. Thus, in order to obtain and investigate stream functions $\psi(\lambda, \theta)$ possessing features similar to those encountered in the incompressible fluid case, we have to introduce and study multivalued solutions of (1.5) possessing singularities similar to those mentioned above.

2. Integral operators for generating stream functions of subsonic flow patterns. Integral operators introduced recently in the theory of partial differential equations [1] yield a successful tool to generate and investigate functions of the kind

[2] Note that the image of the flow pattern covers twice the (schlicht) u, v- and the λ,θ-planes. The point P (the image of $z = \infty$) is a branchpoint. The corresponding streamlines are indicated by the same type of marking.

mentioned above. An integral operator is an operator which transforms a class of functions, say $\mathfrak{A}$, into the solutions of a given partial differential equation

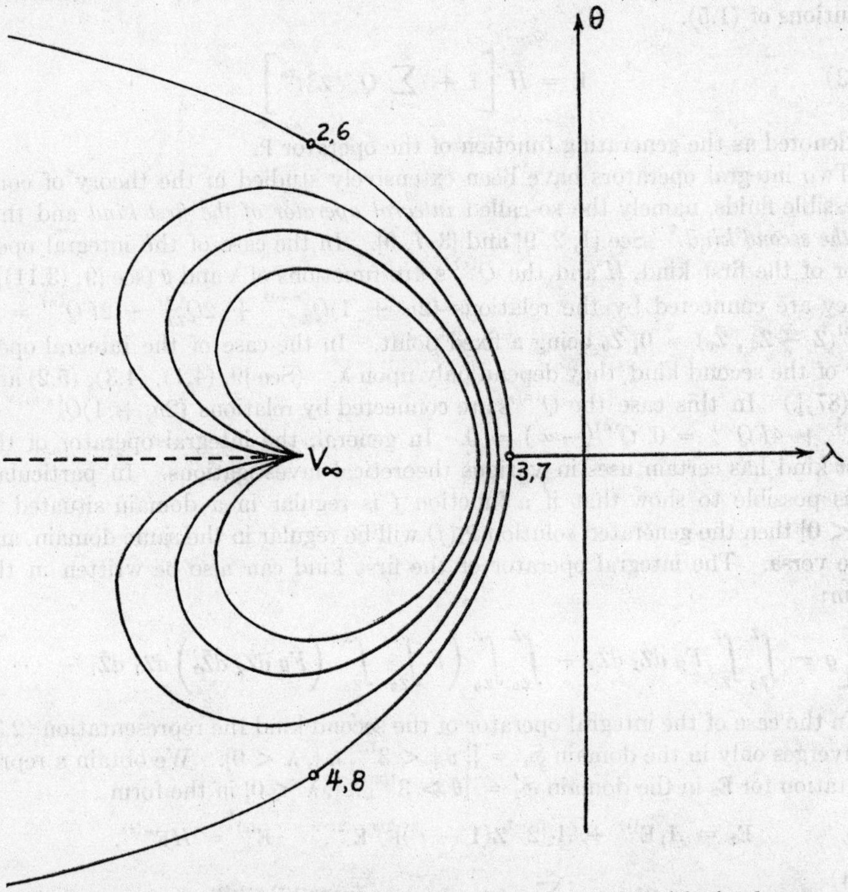

Fig. 1c. The image of a flow around a cylinder in the logarithmic plane

which is, in our case, (1.5). In the subsonic case we use as $\mathfrak{A}$ the class of analytic functions of one complex variable $Z = \lambda + i\theta$ (or $\bar{Z} = \lambda - i\theta$). The integral operator can be written either in the form

$$\text{(2.1)} \qquad P(f) = \int_{t=-1}^{t=1} E(Z, \bar{Z}, t) f\left(\frac{1}{2} Z(1 - t^2)\right) dt \Big/ (1 - t^2)^{1/2}$$

or

$$\text{(2.2)} \qquad p(g) = H\left[g(Z) + \sum_{n=1}^{\infty} \frac{\Gamma(2n + 1)Q^{(n)}}{2^{2n}\Gamma(n + 1)} \int_0^Z \int_0^{Z_1} \cdots \int_0^{Z_{n-1}} g(Z_n) dZ_n \cdots dZ_1 \right]$$

$$= H\left[g(Z) + \sum_{n=1}^{\infty} \frac{(-1)^{n-1} Q^{(n)}}{2^{2n} B(n, n + 1)} \int_0^Z (Z - \zeta)^{n-1} g(\zeta)\, d\zeta \right],$$

where f and g are arbitrary analytic functions of one complex variable and $E(Z, \bar{Z}, t)$, H and the $Q^{(n)}$'s are certain fixed functions. Functions f and g are denoted as the P-associate and the p-associate,[3] respectively, of the generated solutions of (1.5).

$$E = H\left[1 + \sum_{n=1}^{\infty} Q^{(n)} Z^n t^{2n}\right] \tag{2.3}$$

is denoted as the generating function of the operator P.

Two integral operators have been extensively studied in the theory of compressible fluids, namely the so-called *integral operator of the first kind* and that of *the second kind.*[4] See [1, 2, 9] and [3–7, 9]. In the case of the integral operator of the first kind, H and the $Q^{(n)}$'s are functions of λ and θ (see [9; (3.11)]). They are connected by the relations $(2n + 1)Q_{\bar{Z}}^{(n+1)} + 2Q_{Z\bar{Z}}^{(n)} + 2FQ^{(n)} = 0$, $Q^{(n)}(Z - Z_0, \bar{Z}_0) = 0$, Z_0 being a fixed point. In the case of the integral opertor of the second kind, they depend only upon λ. (See [9, (4.1), (4.3), (5.2) and 4, (87)].) In this case the $Q^{(n)}$'s are connected by relations $(2n + 1)Q_{\lambda}^{(n+1)} + Q_{\lambda\lambda}^{(n)} + 4FQ^{(n)} = 0$, $Q^{(n)}(-\infty) = 0$. In general, the integral operator of the first kind has certain uses in various theoretical investigations. In particular, it is possible to show that if a function f is regular in a domain situated in $[\lambda < 0]$ then the generated solution $P_1(f)$ will be regular in the same domain, and vice versa. The integral operator of the first kind can also be written in the form:

$$H\left[g - \int_{Z_0}^{Z}\int_{\bar{Z}_0}^{\bar{Z}} Fg\, dZ_1\, d\bar{Z}_1 + \int_{Z_0}^{Z}\int_{\bar{Z}_0}^{\bar{Z}} \left(F \int_{Z_0}^{Z_1}\int_{\bar{Z}_0}^{\bar{Z}_1} \left(Fg\, dZ_2\, d\bar{Z}_2\right) dZ_1\, d\bar{Z}_1 - \cdots\right]\right.$$

In the case of the integral operator of the second kind the representation (2.3) converges only in the domain $\mathfrak{S}_1 = [|\,\theta\,| < 3^{1/2}|\,\lambda\,|, \lambda < 0]$. We obtain a representation for E_2 in the domain $\mathfrak{S}_2' = [\theta > 3^{1/2}|\,\lambda\,|, \lambda < 0]$ in the form

$$E_2 = A_1 E^{(1)} + A_2[2^{-1}Z(1 - t^2)]^{2/3} E^{(2)}, \qquad E^{(\kappa)} = HE^{*(\kappa)},$$
$$E^{*(\kappa)} = \sum_{n=0}^{\infty} q^{(n,\kappa)}\,(2\lambda)/(-t^2 Z)^{n-(1/2)+(2\kappa/3)}, \qquad \kappa = 1, 2, \tag{2.4}$$

---

[3] f and g are connected by the relation: $f(Z) = (2Z)^{1/2} d^{1/2} g(2Z)/\Gamma(1/2)\, dZ^{1/2} = (1/\pi)g(0) + (2/\pi)\int_0^{\pi/2} Z \sin\vartheta (dg(2Z\sin^2\vartheta)/d(Z\sin^2\vartheta))\,d\vartheta$. In some instances it is more convenient to use operators in the form (2.1); in other instances in the form (2.2).

[4] Usually we denote by $P_\kappa(f)$ the integral operator of the κth kind, $\kappa = 1, 2$. We remark that in [3–7] we have not explicitly indicated that the operator P_2 (and p_2), which have been there denoted by P (and p, respectively) is the integral operator of the *second* kind.

Integral operators transform analytic functions of one complex variable into complex solutions of the compressibility equation. If we change the integral operator, that is, change the generating function, the relations between $\mathrm{Re}[P(f)]$ and $\mathrm{Im}[P(f)]$ will be changed.

The integral operator of the first kind has in the case of [1.7] the following important property. If we continue the values of $\psi = \mathrm{Re}[P(f)]$ and of $P(f)$ to complex values of the arguments λ and θ, that is, we assume that Z and Z are two independent variables, then these continuations in the characteristic planes $Z = 0$ and $\bar{Z} = 0$, respectively, will be functions which differ only by a constant.

where the $q^{(n,\kappa)}(2\lambda)$, $\kappa = 1, 2$; $n = 0, 1, 2, \cdots$ (which for $|\lambda| < (h^{-1} - 1)\pi/2$ can be represented in the form

$$\sum_{\nu=0}^{\infty} C_\nu^{(n,\kappa)}(-\lambda)^{n-(1/2)+(2/3)(\kappa+\nu)})$$

are functions which are connected by equations

$$2(n + 2\kappa/3)q_\lambda^{(n,\kappa)} + q_{\lambda\lambda}^{(n+1,\kappa)} + 4Fq^{(n+1,\kappa)} = 0.$$

REMARK 2.1. We note that if we replace, in the expressions for $E^{*(\kappa)}$, $-\lambda$ by $i\Lambda$ (see 1.10), then E_2 will be defined in the supersonic region $\mathfrak{S}_2'' = [\theta > \Lambda, \Lambda > 0]$. If further we introduce the variable s, given by $s = (-\lambda)^{2/3}$ for $M < 1$, $s = -\Lambda^{2/3}$ for $M > 1$, see [9, (4.23)], then functions $\psi = \int_{t=-1}^{1} E_2 f \, dt/(1 - t^2)^{1/2}$, where E_2 is given by (2.4), are analytic in s and θ (and therefore in M and θ) in $\mathfrak{S}_2 = \mathfrak{S}_2' + \mathfrak{S}_2''$, see [9, (4.24)]. Thus E_2 given by (2.4) can be used to generate partially subsonic and partially supersonic flow patterns which are defined in $\mathfrak{S}_2$. E_2 can also be used for generating "mixed" flow patterns in the corresponding domain situated in the lower half plane, $\theta < 0$. See [9, §5].

Since computation and tabulation of functions which depend on *two* variables is very cumbersome, we use for numerical purposes as a rule the integral operator of the second kind, despite certain theoretical difficulties which arise in applying it. Tables of the first six $Q^{(n)}$'s and of H for the integral operator of the second kind can be found in [12]. The functions $q^{(n,\kappa)}$ can be tabulated in a similar manner.

Suppose $g(Z)$ is the complex potential (in the logarithmic plane) which in the physical plane yields a flow (of an incompressible fluid) around a profile, say $\mathfrak{B}$. Let $\psi = \operatorname{Im}[p_2(g)]$. If certain conditions (see condition I, p. 27) concerning the behavior at singular points of g are fulfilled, then by using the formula

$$x = \int \rho^{-1}\{[-v^{-2}(1 - M^2)\cos\theta\,\psi_\theta - v^{-1}\sin\theta\psi_v]\,dv + [\cos\theta\psi_v - v^{-1}\sin\theta\,\psi_\theta]\}\,d\theta,$$

$$(2.5)\qquad y = \int \rho^{-1}\{[-v^{-2}(1 - M^2)\sin\theta\,\psi_\theta + v^{-1}\cos\theta\,\psi_v]\,dv + [\sin\theta\psi_v + v^{-1}\cos\theta\,\psi_\theta]\}\,d\theta$$

[4, §14], we can determine the corresponding flow pattern in the physical plane, which will represent a flow of a *compressible* fluid around a slightly distorted profile $\mathfrak{B}$. See Figs. 1a, 1b, 1c.

The integral operators of the first and second kinds generate stream functions possessing in most cases the same general features as the functions to which they are applied. In particular, if the function g has a branch point of nth order, the generated solution of (1.5) will have a branch point of the same order and at the same point as g. There exists, however, an important exceptional case which requires a modification in employing this procedure in hydrodynamical applications; namely if we apply the operator p to a function g possessing a pole or a logarithmic singularity, then the operator produces an essential modification of the behavior of the solution at the aforementioned singular point. Namely *both* the real and imaginary parts of $p[\lg(Z - Z_\kappa)]$ and $p[(Z - Z_\kappa)^{-n}]$, n a positive integer, will be infinitely many-valued at $Z = Z_\kappa$. Since, in connection with the transition to the physical plane, it is of importance to have singularities for which at least one of the functions φ or ψ is single-valued at the singular point, it is necessary to modify our operator in the following manner. If we have a function g which has a pole or logarithmic singularity we decompose it into two parts, the first, say g_1, having neither poles nor logarithmic singularities, and the second being

$$\sum_{\kappa=1}^{m} A_\kappa (Z - Z_\kappa)^{-\nu_\kappa} + \sum_{\kappa=m}^{n} A_\kappa \lg(Z - Z_\kappa), \tag{2.6}$$

where A_κ are constants and the ν_κ are positive integers.

Applying the integral operator to the first part, we obtain a solution of (1.5), say $\psi_1 = \mathrm{Im}[P(g)]$.

By introducing so-called *fundamental solutions*

$$\psi^{(L)}(Z; Z_\kappa) = A(Z; Z_\kappa) \log | Z - Z_\kappa | + B(Z; Z_\kappa) \tag{2.7}$$

of (1.5) or (1.9), where A and B are certain single-valued functions[5] which can be easily determined, we obtain solutions of (1.5) [or (1.9)] which have the same behavior as $\log | Z - Z_k |$. See [4, p. 45; 6, p. 43; 7]. Differentiating these functions with respect to the real and imaginary parts of the parameter Z_κ we obtain single-valued functions $\psi^{(1,1)}$ and $\psi^{(1,2)}$ with an infinity of the first order.

Remark 2.2. Flow patterns generated by $\psi^{(L)}$, $\psi^{(1,1)}$, $\psi^{(1,2)}$ have a simple physical meaning. $\psi^{(L)}$ yields a vortex at infinity (in the physical plane), a fundamental solution $\varphi^{(L)}$ of (1.9) yields a source or a sink at infinity. $\psi^{(1,1)}$, $\psi^{(1,2)}$ yield components of a doublet at infinity.

Adding to ψ_1 (see line 15 of this page) the functions with singularities described above we obtain a solution of the compressibility equation, possessing the same features as g. We shall refer to this procedure for determining stream functions of a compressible fluid as *the modified integral operator*.

In the case of non-symmetric flow patterns with the velocity vector $V_0 e^{i\theta_0}$ at infinity ($z = \infty$), we obtain as a rule the stream function (in the λ, θ-plane) in the form

$$\psi = A_0 \psi^{(L)}(Z; Z_0) + A_1 \psi^{(1,1)}(Z; Z_0) + A_2 \psi^{(1,2)}(Z; Z_0) + \psi_1(Z) \tag{2.8}$$

where $Z_0 = \lambda(v_0) + i\theta_0$, and $\psi_1(Z)$ is a function which has branch points of positive order as its only singularities, and which can be obtained by the operator from a suitably chosen function $[g(Z) - g_1(Z)]$ of one complex variable.

---

[5] We note that in general A and B are *not* analytic functions of one complex variable Z (or Z_κ). In order to emphasize this they are often denoted $A(Z, \bar{Z}; Z_\kappa, \bar{Z}_\kappa)$, $B(Z, \bar{Z}; Z_\kappa, \bar{Z}_\kappa)$ respectively.

Concerning another method for determining a fundamental solution, see [15].

In the case of symmetric flow patterns the representation (2.8) simplifies. In the latter case we can directly apply the p_2 operator to the function

$$g(Z) = \sum_{\nu=1}^{\infty} A_\nu (Z - Z_0)^{\nu/2}. \tag{2.9}$$

See [5; 11].

The introduction of modified integral operators enables us to develop a theory of compressible fluid flows which in the subsonic case strongly resembles that of an incompressible fluid (when we use the hodograph method). In the following, a generalization of two results of the latter theory will be discussed.

I. *Conditions that a stream function defined in the* λ, θ*-plane yields a flow pattern in the physical plane around a closed curve.*

In the compressible fluid case, (2.5) can be written in the form

$$z = \int \frac{dw}{\exp Z^{\blacktriangle}} = \int \frac{1}{\exp Z^{\blacktriangle}} \cdot \frac{dw}{dZ^{\blacktriangle}}\, dZ^{\blacktriangle}, \tag{2.10}$$

where $w = \phi + i\psi$, $Z^{\blacktriangle} = \lambda^{\blacktriangle} - i\theta$, $\lambda^{\blacktriangle} = \log v$.

Taking then the real and the imaginary parts of the right-hand side of (2.10) and using the Cauchy-Riemann equations connecting ϕ and ψ, we obtain in the incompressible fluid case for x and y formulae analogous to (2.5). On the other hand if the domain of definition of a stream function $\mathfrak{B}$ (in the λ,θ-plane) is simply connected, then the necessary and sufficient condition that the boundary curve of the image of $\mathfrak{B}$ in the x,y-plane is a closed curve is that the sum of the residues of the singularities $[(\exp Z^{\blacktriangle})^{-1}(dw/dZ^{\blacktriangle})]$ situated in $\mathfrak{B}$ equals[6] 0.

This result can be generalized to the compressible fluid case. The integrands of (2.5) are complete differentials, so the values of these integrals are independent of the path of integration, provided that no singular point is crossed. If therefore we substitute into (2.5) the fundamental solution (2.7) (or their derivatives with respect to Re Z_0 or to Im Z_0), we can associate with each such a singularity two real numbers,

$$T^{(1)}(Z_0) \quad \text{and} \quad T^{(2)}(Z_0). \tag{2.11}$$

(In the incompressible fluid case these numbers are the real and the imaginary parts of the residue at Z_0.) If the domain $\mathfrak{B}$ (with boundary curve b) is simply-connected, then the necessary and sufficient condition that the image of b in the x, y-plane is a closed curve is that

$$\sum_{\nu=0}^{2} A_\nu T_\nu^{(s)}(Z_0) = 0, \qquad s = 1, 2, \tag{2.12}$$

where A_ν are the coefficients appearing in (2.8) [4, §14; 6; 8, §6].

II. *A generalization of Blasius formulae.* A further application of our method

[6] In the case of multiply-connected domains we have in addition to this to take into consideration the so-called periods of the integrals.

of attack is a generalization of the formulae of Blasius expressing the lift $\boldsymbol{F}$ and moment M exerted upon an immersed body, $\mathfrak{B}$, in quantities characteristic for the behavior of the flow at infinity, $z = \infty$. Let us assume that a profile $\mathfrak{B}$ is immersed in a flow of a compressible fluid, and the streamfunction (in the λ,θ-plane) has the form (2.8). For the force $\boldsymbol{F}$ exerting upon the immersed body $\mathfrak{B}$ we obtain the expression

$$(2.13) \quad \boldsymbol{F} = \mathcal{F}(\psi) = -\int_{\mathcal{C}} v(\lambda)e^{i\theta}\, d\psi + i\sigma \int_{\mathcal{C}} [1 - 2^{-1}(k-1)(v^2(\lambda))]^{k/(k-1)}\, dz$$

where

$$(2.14) \quad dz = \frac{1}{\rho} e^{i\theta}\left\{\left[-\frac{1-M^2}{v^2}\psi_\theta\left(\frac{dv}{d\lambda}\right) + i\frac{\psi_\lambda}{v}\right]d\lambda + \left[\psi_\lambda\left(\frac{dv}{d\lambda}\right)^{-1} + i\frac{\psi_\theta}{v}\right]d\theta\right\}$$

and $\mathcal{C}$ is any sufficiently smooth and sufficiently small closed curve around Z_0. In generalization of the Blasius formulae we obtain in the compressible fluid case

$$(2.15) \qquad \boldsymbol{F} = A_0\mathcal{F}(\psi^{(L)}) + A_1\mathcal{F}(\psi^{(1,1)}) + A_2\mathcal{F}(\psi^{(1,2)})$$

where $\mathcal{F}(\psi^{(L)}), \cdots$ are vectors which we obtain by substituting $\psi^{(L)}, \cdots$ in (2.13), instead of ψ. An analogous formula can be obtained for the moment, M [8; §6].

3. Mixed flow patterns. In §2 we have considered only the subsonic case. The theory of integral operators can in an analogous way be applied to get purely supersonic flow patterns.[7] In this case we chose as class $\mathfrak{A}$ to which the operator is applied the totality of twice differentiable functions of one real variable s (or r) where

$$(3.1) \quad \mathrm{s} = \Lambda - \theta, \mathrm{r} = \Lambda + \theta, \Lambda = -\arctan(M^2-1)^{1/2} + h^{-1}\arctan(h(M^2-1)^{1/2}).$$

A certain handicap in this approach is the fact that the theory of single- and multivalued solutions of $\psi_{\theta\theta} - \psi_{\Lambda\Lambda} = 0$ is not sufficiently explored from the point of view of the structure of their level lines. In particular, in contrast to the subsonic case, the stream functions yielding in the physical plane flow patterns around profiles of interest in physical applications have not yet been determined in the case of the latter equation. On the other hand, it seems that this method of attack would not be of considerable interest in applications since most flows in which we are interested are mixed, that is, partially subsonic and partially supersonic, and in addition may have shock waves. This complication makes necessary a considerable modification in the approach.

In this section, we shall consider the mixed case, without shock waves. As in the subsonic case, it is useful at first to develop the theory of flows governed by a simplified equation, that is, an equation whose solutions have a behavior similar to that of solutions of the "exact" compressibility equation but whose mathematical theory is considerably simpler. By employing suitably chosen operators one can then transform solutions of the simplified equation into solutions of the exact one.

[7] It should be stressed here that this generalization does not introduce any additional difficulty.

I. *The case of simplified compressibility equation.* (See [20, part II]). We obtain a simplified equation of this type if we replace in equation (1.2) the coefficient $l(\mathrm{H})$, whose series development is given by (1.3), by its first term, that is, if we replace $l(\mathrm{H})$ by

$$l^\dagger(\mathrm{H}^\dagger) = -C\mathrm{H}^\dagger, \quad C = 2[2/(k-1)]^{(2-k)/(k-1)} \tag{3.2}$$

[7, 8, 9]. We obtain then

$$S^\dagger(\psi) = -C\mathrm{H}^\dagger\psi_{\theta\theta} + \psi_{\mathrm{H}^\dagger\mathrm{H}^\dagger} = 0, \qquad C = \text{constant}. \tag{3.3}$$

If by the transformation

$$-\mathrm{H}^\dagger = c(-\lambda^\dagger)^{2/3} \tag{3.4}$$

we reduce this equation to normal form then we obtain

$$(\partial^2\psi^\dagger/\partial\lambda^{\dagger 2}) + (\partial^2\psi^\dagger/\partial\theta^2) + (3/\lambda^\dagger)^{-1}(\partial\psi^\dagger/\partial\lambda^\dagger) = 0 \tag{3.5}$$

or [introducing

$$\psi^{*\dagger} = \psi^\dagger/(-2\lambda^\dagger)^{1/6}\ ] \tag{3.6}$$

$$(\partial^2\psi^{*\dagger}/\partial\lambda^{\dagger 2}) + (\partial^2\psi^{*\dagger}/\partial\theta^2) + (5/36\lambda^{\dagger 2})\psi^{*\dagger} = 0. \tag{3.7}$$

Remark 3.1. Instead of simplifying the equation $S(\psi) = 0$, we can simplify equation (1.7) by replacing F by the first term of its series development at $\lambda = 0$. We then obtain equation (3.7) once again. We note further that it is *independent* of k.

If we replace the function $4F$ by $4F^\dagger = 5/36\lambda^{\dagger 2}$ then the expressions for the generating functions given in (2.3) and (2.4)[8] essentially become hypergeometric series, namely

$$\begin{aligned} \mathrm{E}_2^\dagger = {} & A_1(-2\lambda^\dagger)^{-1/6}\mathrm{F}\left(\frac{1}{6}, \frac{5}{6}, \frac{1}{2}, \frac{-t^2(\lambda^\dagger + i\theta)}{-2\lambda^\dagger}\right) \\ & + A_2(-2\lambda^\dagger)^{-2/3}[-t^2(\lambda^\dagger + i\theta)]^{1/2}\,\mathrm{F}\left(\frac{2}{3}, \frac{4}{3}, \frac{3}{2}, \frac{-t^2(\lambda^\dagger + i\theta)}{-2\lambda^\dagger}\right) \end{aligned} \tag{3.8}$$

for $|(\lambda^\dagger + i\theta)/-2\lambda^\dagger| < 1$, and

$$\begin{aligned} \mathrm{E}_2^\dagger = {} & A_3[-t^2(\lambda^\dagger + i\theta)]^{-1/6}\mathrm{F}\left(\frac{1}{6}, \frac{2}{3}, \frac{1}{3}, \frac{-2\lambda^\dagger}{-t^2(\lambda^\dagger + i\theta)}\right) \\ & + A_4(-2\lambda^\dagger)^{2/3}[-t^2(\lambda^\dagger + i\theta)]^{-5/6}\,\mathrm{F}\left(\frac{5}{6}, \frac{4}{3}, \frac{5}{3}, \frac{-2\lambda^\dagger}{-t^2(\lambda^\dagger + i\theta)}\right) \end{aligned} \tag{3.9}$$

for $|-2\lambda^\dagger/(\lambda^\dagger + i\theta)| < 1$. The representation can be extended to the case $|(-2\lambda^\dagger)/(\lambda^\dagger + i\theta)| = 1$. One of the main problems which arises in the mixed

[8] We note that H and $\mathrm{H}^\dagger = c(-2\lambda^\dagger)^{-1/6}$ differ. We note that the second term in (3.8) can be neglected since $\int_{t=-1}^{1} t^{-1}(-\lambda^\dagger - i\theta)^{-1/2}\mathrm{F}(2/3,\ 4/3,\ 3/2,\ -t^2(\lambda^\dagger + i\theta)/-2\lambda^\dagger) \cdot f(2^{-1}(1-t^2)\mathrm{Z}^\dagger)dt/(1-t^2)^{1/2} = 0$ if $f(\zeta)$ is regular in the neighborhood of $\zeta = 0$ and vanishes at $\zeta = 0$. This is the reason it was not indicated in §2.

case is to determine conditions on the associate function f in order that the corresponding solution of the compressibility equation together with its first derivative with respect to H (and therefore also to M) will be continuous on the transonic line (H = 0, that is, $M = 1$). Conditions sufficient to assume this can be given in various forms. For instance, sufficient conditions to this effect are that in the neighborhood of the point $Z^\dagger = 0$, that is, $\lambda^\dagger = 0, \theta = 0, f(Z^\dagger)$ and $f'(Z^\dagger)$ may be approximated arbitrarily closely by $\Sigma_{\nu=1}^{N} a_\nu^{(N)} Z^{\dagger n\nu}$, $n \geqq 5/6$, that is, that for every $\epsilon > 0$ constants $a_\nu^{(N)}$ can be determined such that

$$(3.10) \qquad |f(Z^\dagger) - \sum_{\nu=1}^{N} a_\nu^{(N)} Z^{\dagger n_\nu}| < \epsilon, \qquad |f'(Z^\dagger) - \sum_{\nu=1}^{N} n_\nu a_\nu^{(N)} Z^{\dagger n_\nu - 1}| < \epsilon$$

for $Z^\dagger$ belonging to the domain of definition of the flow pattern. See [9; §4].

Our approach permits of various modifications. So in the case of symmetric profiles the following procedure for constructing mixed flows from a large class of analytic functions can be derived. The stream function of an incompressible flow around a symmetrical obstacle in the physical plane in the case under consideration can be written in the logarithmic plane in the form

$$(3.11) \qquad \psi^\blacktriangle = \operatorname{Im}\left[\sum_{\nu=0}^{\infty} b_\nu (Z^\blacktriangle - Z_0)^{\nu - 1/2}\right]$$

where $Z_0 = \log v_0 - i\theta_0$, $v_0 \exp(i\theta_0)$ being the velocity vector at infinity, $z = \infty$. Replacing $Z^\blacktriangle$ by $Z^\dagger$, applying the integral operator of the second kind and carrying out certain additional transformations, we obtain for the solution of the simplified compressibility equation (3.5) the expressions

$$(3.12) \qquad \psi^\dagger = \operatorname{Im}\left[\sum_{\nu=0}^{\infty} b_\nu (Z^\dagger - Z_0)^{\nu-1/2} F\left(\frac{1}{6}, \frac{5}{6}, \nu + \frac{1}{2}, \frac{Z^\dagger - Z_0}{2\lambda^\dagger}\right)\right],$$

which then gives in the physical plane a flow pattern of a compressible fluid around a profile similar to the profile $\mathfrak{B}$. In this case, no additional conditions need be imposed on the associate function to assure the aforementioned behavior on H = 0. See [9, §4; 11].

II. *The case of exact compressibility equation.* **1.** 1. If we replace in (2.1) $i\lambda$ by Λ then in the case of the integral operator of the first kind the expressions obtained will represent a streamfunction of a supersonic flow pattern. If the associate is regular in a sufficiently large domain, the expressions generated by the integral operator of the first kind are defined in the domain $\mathfrak{T}_1 = E[M < 1]$ and $\mathfrak{T}_2 = E[M > 1]$, respectively. On the other hand, in general the above expressions are not defined along the transsonic line, and additional investigations are needed to obtain mixed flow patterns in this manner.

2. If we use the operator of the second kind then (in analogy to the representations (3.8) for the simplified case) the expressions [4, (87)] and [7, (51)] hold for the regions $\mathfrak{S}_1 = E[3^{1/2}\lambda < \theta < -3^{1/2}\lambda, \lambda < 0]$ and $\mathfrak{S}_4 = E[-3\Lambda < \theta < \Lambda, \Lambda > 0]$, respectively. As has been mentioned in Remark 2.1, the formula

$$(3.13) \qquad \psi = \operatorname{Im}\left\{\int_{t=-1}^{1} E_2(Z, \bar{Z}, t) f(2^{-1}Z(1 - t^2))\, dt/(1 - t^2)^{1/2}\right\},$$

where $E_2 = A_1E^{(1)} + [2^{-1}Z(1 - t^2)]^{2/3}A_2E^{(2)}$, A_κ are constants, $E^{(\kappa)}$ are given by (2.4), and f is a function of one variable, yields a flow pattern defined in $\mathfrak{S}_2 = \mathfrak{S}_2' + \mathfrak{S}_2''$ (see Remark 2.1), as well as in a corresponding domain $\mathfrak{S}_3$ in the lower half-plane $\theta < 0$.[9]

If we introduce new variables $u = t^2Z/(Z + \bar{Z})$ and $s = [-(Z + \bar{Z})/2]^{2/3}$ then the equation

$$2E^*_{\tau\lambda} + E^*_{\lambda\lambda} - \tau^{-1}E_\lambda{}^* + 4FE^* = 0, \tau = t^2Z,$$

for the generating function $E^* = E_2/H$ of the second kind, assumes the form

$$\left[u\left(1 - u\right)E^*_{uu} + \left(\frac{1}{2} - 2u\right)E^*_u - \frac{5}{36}E^*\right]$$

$$+\left[\frac{2}{3}s(2u - 1)E^*_{us} + 9^{-1}s(2 + 3u^{-1})E^*_s - (4/9)s^2E^*_{ss} - 4(S(s) - 5/144)E^*\right] = 0$$

see [9, (4.2) and (4.7)]. Writing then

$$E^* = \sum_{n=0}^{\infty} s^n U^{(n)}(u), \qquad U^{(0)} = E^{*\dagger}, \qquad U^{(1)} = 0,$$

we obtain for the $U^{(n)}$ the recursion formulas,

$$u(1 - u)U^{(n)}_{uu} + \left[\frac{1}{2} - \frac{2n}{3} - \left(2 - \frac{4n}{3}\right)u\right]U^{(n)}_u$$

$$+\left[\frac{2n}{3u} - \frac{(4n - 5)(4n - 1)}{36}\right]U^{(n)} = 4\sum_{\gamma=0}^{n-2}\alpha_{n-\gamma}U^{(\gamma)}.$$

In addition to expressions of the form (2.3) and (2.4), further representations for E^* can be obtained from the above formulae similar to those obtainable in the simplified case by using the theory of the hypergeometric equation.

3. The inversion formulas, that is, formulas expressing the associate f [or g] in terms of the stream function ψ are of considerable interest in various applications of the method of integral operators. In the case of an integral operator of the first kind, we have

$$\psi(Z, 0) = -2^{-1}i[g(Z) - \bar{R}_1(0, Z)\bar{g}(0)].$$

In the case of the integral operator of the second kind similar inversion formulas can be obtained. (See Trans. Amer. Math. Soc. vol. 57 (1945) p. 302 ff.) The above formulas, however, require the knowledge of the analytic continuation of

[9] Integral formulas indicated above yield solutions of (1.2) (that is, stream functions of compressible fluid flows) which are defined in the domains $\mathfrak{T}_1$ and $\mathfrak{T}_2$ respectively (in the case of the integral operator of the first kind) and in $\mathfrak{S}_1$, $\mathfrak{S}_2$, $\mathfrak{S}_3$, $\mathfrak{S}_4$, respectively (in the case of the integral operator of the second kind). The question how to determine the associates g (or f) such that g (or f), inserted in formulas defined in different parts of the plane, yield the analytic continuation of the same stream function ψ can be reduced by the use of inversion formulas to a problem in the theory of functions of a complex variable, see [9, §5], or handled by procedures indicated in §3.4 (see p. 36) of the present paper.

ψ to the complex values of the arguments λ, θ. In the case of the integral operator of the second kind a formula expressing f in terms of the values

$$\lim_{M\to 1^-} \psi(M, \theta) = \chi_1(\theta) \equiv \sum_{\nu=0}^{\infty} a_\nu^{(1)} \theta^\nu,$$

$$\lim_{M\to 1^-} (\partial\psi(M, \theta)/\partial M) = \chi_2(\theta) \equiv 3^{1/3}(1 - h^2)^{2/3} \sum_{\nu=0}^{\infty} a_\nu^{(2)} \theta^\nu$$

can be derived.

Assuming that in (3.13) $\operatorname{Im}\{A_1 A_2\} \neq 0$, we have

$$f(\zeta) = -(-2i\zeta)^{1/6} 3^{-1/2} \pi^{-1} S_0^{-2} [\operatorname{Im}(A_2 \bar{A}_1)]^{-1} \cdot \left[-\bar{D}_0 \int_{\mathcal{C}} \left(t^{-1/3} \chi_1(\sigma) + \sum_{\kappa=1}^{2} (-1)^\kappa \bar{D}_\kappa t^{-5/3} \chi_\kappa(\sigma) \right) dt \right]$$

where $\sigma = -2i\zeta(1 - t^2)$, the constants D_ν, $\nu = 0, 1, 2$, are:

$$D_0 = -2 \cdot 3^{-1} i^{3/2} S_0 A_2, \quad D_1 = -2^{5/3} 3^{-1} i^{1/6} S_0 S_1 A_1, \quad D_2 = -i^{1/6} S_0 A_1,$$

and the S_ν are defined in (1.6). $\mathcal{C}$ is a simple curve in the complex t-plane which connects $t = -1$ with $t = 1$ and, except for the end points, lies outside $E[\,|t| \leqq 1]$. Moreover, $\mathcal{C}$ has to be chosen in such a manner that $[2^{-1}Z(1 - t^2)]$ lies in the regularity domain of f for the values of Z under consideration [9, §6].

2. We wish now to indicate another procedure which aims at constructing "mixed" flows. In applying this method, we use in addition to the integral operator the Chaplygin solutions. Suppose that $g(Z^\blacktriangle)$ is a complex potential which in the physical plane yields a flow past a closed curve. For simplicity's sake let us assume that the flow is symmetric, and the complex potential (in the logarithmic plane) is a two-valued function, possessing as its only singularity a branch point of the second order at the point $\lambda^\blacktriangle = \lambda_\infty$, $\theta = 0$. Employing the operator $\mathbf{p}$ [4, (69), (89)], or that of the first kind, we obtain a solution of equation (1.5), $\psi^{(1)} = \operatorname{Im}[\mathbf{p}(g)]$. If the maximum speed is chosen sufficiently large, this solution will be determined at first only in a portion, say D, of its domain of definition, namely in the portion situated in $[\lambda \leqq \lambda_0]$, $\lambda_0 < 0$, so that, in general, the boundary curve $\psi^{(1)} = 0$ is not closed in the aforementioned domain. Let us assume that for the λ-coordinate λ_∞ of the branch points $\lambda_\infty < \lambda_0$ holds. We shall now describe a procedure (different from that indicated in **1** and in the footnote 9) to determine the analytic continuation of the solution $\psi^{(1)}$ outside of D. Since in the following we have also to operate in the supersonic region, and as we shall employ Chaplygin's solutions, it is convenient to introduce, instead of λ, the variable $\tau = (k - 1)v^2/2a_0^2$, where a_0 is the speed of sound at a stagnation point. In the subsonic region $\tau = \tau(\lambda)$ is a biuniform function of λ. Let $\tau_\infty = \tau(\lambda_\infty)$. According to our assumptions for $\tau > \tau_\infty$ the streamfunction is defined only in two disconnected regions, say $\mathcal{B}_1$ and $\mathcal{B}_2$, which have no branch points in their interiors. (These domains lie on two different sheets of the Riemann surface.) Let $\tau_0 = \tau(\lambda_0) > \tau_\infty$ and let us now introduce Chaplygin's solutions[10]

$$(3.14) \qquad (A_\nu \mathrm{F}_\nu + B_\nu \mathrm{F}_\nu^*)\tau^{\nu/2L} e^{i(\pi\nu\theta)/L},$$

where

$$(3.15) \qquad \mathrm{F}_\nu = \mathrm{F}(\alpha_\nu, \beta_\nu, -\beta, 1-\tau),$$

$$\mathrm{F}_\nu^* = (1-\tau)^{\beta+1}\mathrm{F}(\gamma_\nu - \alpha_\nu, \gamma_\nu - \beta_\nu; 2+\beta, 1-\tau),$$

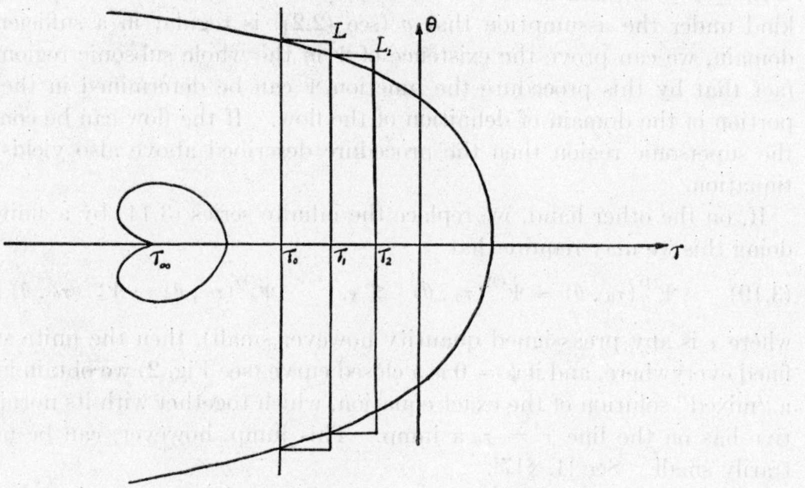

Fig. 2

F being the hypergeometric series. Here

$$(3.16) \qquad \begin{aligned} \gamma_\nu &= \left(\frac{\nu}{L} + 1\right), \quad \beta = \frac{1}{k-1}, \quad \alpha_\nu = 2^{-1}\left[\left(\frac{\nu}{L} - \beta\right) + \Delta_\nu\right], \\ \beta_\nu &= 2^{-1}\left[\left(\frac{\nu}{L} - \beta\right) - \Delta_\nu\right], \quad \Delta_\nu = \left[\left(\frac{\nu}{2L}\right)^2 (2\beta+1) + \frac{1}{4}\beta^2\right]^{1/2}. \end{aligned}$$

In order to compute the continuation of $\Psi^{(1)}(\tau, \theta) = \psi^{(1)}(\lambda(\tau), \theta)$, obtained in D by the use of the integral operator into the domain $\mathfrak{B}_1$ (and $\mathfrak{B}_2$ respectively) we write $\Psi^{(2)}$ in the form

$$(3.17) \quad \Psi^{(2)} = \sum_{\nu=1}^{\infty} (A_\nu^{(1)}\mathrm{F}_\nu + B_\nu^{(1)}\mathrm{F}_\nu^*)\tau^{\nu/2L} \cos\frac{\pi\nu\theta}{L} + (A_\nu^{(2)}\mathrm{F}_\nu + B_\nu^{(2)}\mathrm{F}_\nu^*)\tau^{\nu/2L} \sin\frac{\pi\nu\theta}{L}$$

[10] The expression $(A_\nu \mathrm{F}_\nu + B_\nu \mathrm{F}_\nu^*)$ which appears in (176) of [4], see also (3.14), depending as it does upon two independent constants, is the most general solution of the hypergeometric equation (175) of [4]; consequently it would be of interest if the authors of [21] would clarify their statement (see in particular p. 4, lines 14–21 of [21]) indicating in which way the author of [4] has restricted his attention to the first solution of the hypergeometric equation alone.

and determine the coefficients $A_\nu^{(\kappa)}$, $B_\nu^{(\kappa)}$, $\kappa = 1, 2$, $\nu = 0, 1, 2, \cdots$, so that on[11] $\tau = \tau_0$, $-L \leqq \theta \leqq L$,

$$\Psi^{(1)}(\tau_0, \theta) = \Psi^{(2)}(\tau_0, \theta), \quad \Psi_\tau^{(1)}(\tau_0, \theta) = \Psi_\tau^{(2)}(\tau_0, \theta), \quad \Psi_\tau^{(1)} \equiv \partial\Psi^{(1)}/\partial\tau, \cdots \tag{3.18}$$

holds. It can be shown that the series (3.17) converges in a certain strip, say $[\tau_0 \leqq \tau \leqq \tau_1, -L \leqq \theta \leqq L]$. By replacing τ_0 by τ_1, L by L_1 and repeating the same procedure we may continue the function to a range of values $[\tau_1 \leqq \tau \leqq \tau_2, -L_1 \leqq \theta \leqq L_1]$ and so on. If we employ the integral operator, $\mathfrak{p}_1$, of the first kind under the assumption that g (see (2.2)) is regular in a sufficiently large domain, we can prove the existence of Ψ in the whole subsonic region, and the fact that by this procedure the function Ψ can be determined in the subsonic portion of the domain of definition of the flow. If the flow can be continued to the supersonic region then the procedure described above also yields the continuation.

If, on the other hand, we replace the infinite series (3.14) by a finite one (in doing this we may require that

$$| \Psi^{(1)}(\tau_0, \theta) - \Psi^{(2)}(\tau_0, \theta) | \leqq \epsilon, \qquad | \Psi_\tau^{(1)}(\tau_0, \theta) - \Psi_\tau^{(2)}(\tau_0, \theta) | \leqq \epsilon \tag{3.19}$$

where ϵ is any preassigned quantity however small), then the finite sum is defined everywhere, and if $\psi = 0$ is a closed curve (see Fig. 2) we obtain in this way a "mixed" solution of the exact equation, which together with its normal derivative has on the line $\tau = \tau_0$ a jump. This jump, however, can be made arbitrarily small. See [4, §17].

3. Let us finally indicate still another method for generating single-valued solutions of compressibility equations which yields solutions different from those of Chaplygin. These solutions, in general, lead to certain types of mixed flow patterns (*not* including flows around closed curves).

In [2, pp. 23–27] and [3, §2] the author of the present paper introduced the following operator to generalize potential functions $\varphi(\mathrm{H}, \theta)$ (see (1.1)) and stream-functions $\psi(\mathrm{H}, \theta)$ of the (exact) compressibility equation. Functions φ and ψ defined by

$$\begin{aligned} \varphi + i\psi = (\theta + i\mathrm{H})^{[n]} &\equiv \Bigg\{ \left[\theta^n - 2!\binom{n}{2}\theta^{n-2} \int_0^{\mathrm{H}} l(\mathrm{H}_2)\, d\mathrm{H}_2 \int_0^{\mathrm{H}_2} d\mathrm{H}_1 + \cdots \right] \\ &+ i\left[1!\binom{n}{1}\theta^{n-1} \int_0^{\mathrm{H}} d\mathrm{H}_1 - 3!\binom{n}{3}\theta^{n-3} \right. \\ &\left. \cdot \int_0^{\mathrm{H}} d\mathrm{H}_3 \int_0^{\mathrm{H}_3} l(\mathrm{H}_2)\, d\mathrm{H}_2 \int_0^{\mathrm{H}_2} d\mathrm{H}_1 + \cdots \right] \Bigg\}, \end{aligned} \tag{3.20}$$

[11] Another possibility of continuing the solution would consist in matching together two different representations along the line $\lambda = \lambda_\infty$, that is, the line on which the branchpoint lies. This method, however, offers considerable mathematical difficulties. The series for ψ cannot converge on the line $\lambda = \lambda_\infty$ since the function is singular on this line. Therefore, it is necessary to employ some summation methods in order to prove that the solution and its derivative converge from both sides of $\lambda = \lambda_\infty$ to the same values. Further, it is necessary to show that the stream function and the potential function obtained in this manner are two-valued (and not, say, infinitely many-valued) functions at the singular point.

$$\varphi + i\psi = i\odot(\theta + i\mathrm{H})^{\{n\}} \equiv - \left\{\left[1!\binom{n}{1}\theta^{n-1}\int_0^{\mathrm{H}} l(\mathrm{H}_1)\,d\mathrm{H}_1\right.\right.$$

$$(3.21) \qquad \left. - 3!\binom{n}{3}\theta^{n-1}\int_0^{\mathrm{H}} l(\mathrm{H}_3)\,d\mathrm{H}_3\int_0^{\mathrm{H}_3} d\mathrm{H}_2\int_0^{\mathrm{H}_2} l(\mathrm{H}_1)\,d\mathrm{H}_1 + \cdots\right]$$

$$\left. - i\left[\theta^n - 2!\binom{n}{2}\theta^{n-2}\int_0^{\mathrm{H}} d\mathrm{H}_2\int_0^{\mathrm{H}_2} l(\mathrm{H}_1)\,d\mathrm{H}_1 + \cdots\right]\right\},$$

see (1.1), (1.3), are, in the special case where $l(\mathrm{H}) = 1$, connected by the Cauchy-Riemann equations, and can be interpreted as the potential and the stream

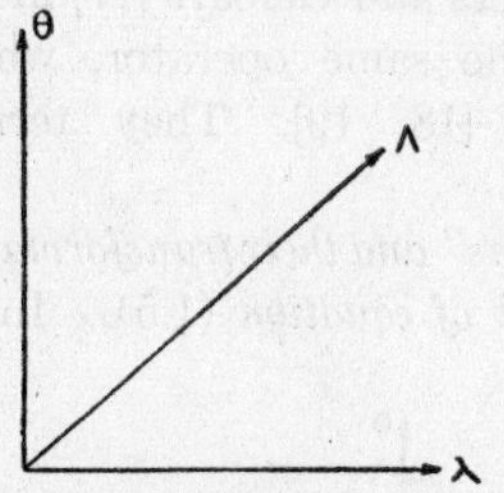

Fig. 3a. Intersection with $\vartheta = -\vartheta_1 < 0$

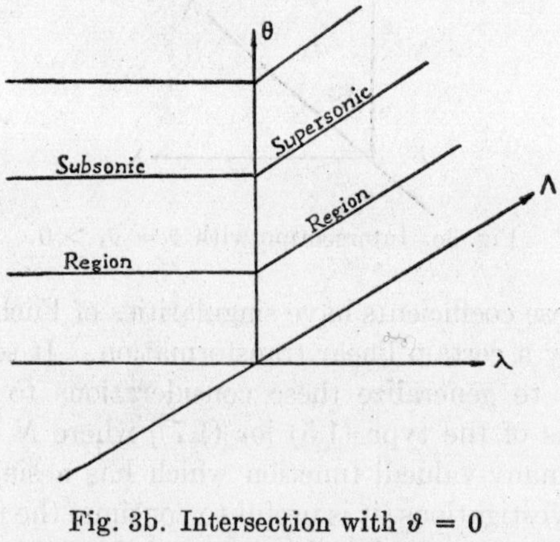

Fig. 3b. Intersection with $\vartheta = 0$

Fig. 3. Successive intersection of the four-dimensional $(\lambda, \Lambda, \theta, \vartheta)$-space, with $\vartheta = \text{const.}$

function of an incompressible fluid flow. In the case of an arbitrary $l(\mathrm{H})$, φ and ψ are connected by the equations

$$(3.22) \qquad \varphi_\theta = \psi_\mathrm{H}, \qquad \varphi_\mathrm{H} = -l(\mathrm{H})\psi_\theta, \qquad \varphi_\theta = \partial\varphi/\partial\theta, \cdots$$

and if we choose $l(\mathrm{H}) = (1 - M^2)/\rho^2$ where M is the local Mach number and $\rho = \rho(M)$ the density, then φ and ψ can be interpreted as the potential and stream function of a *compressible* fluid. For details see [3].

Let us denote, as indicated above, by $(\theta + i\mathrm{H})^{[n]}$ and $i\odot(\theta + i\mathrm{H})^{\{n\}}$ the expression obtained from $(\theta + i\mathrm{H})^n$ by the above procedures. We define, in general, the operation $\boldsymbol{R}_2$ by the relation

$$(3.23) \qquad \boldsymbol{R}_2[\sum(\alpha_n + i\beta_n)(\theta + i\mathrm{H})^n] = \sum[\alpha_n(\theta + i\mathrm{H})^{[n]} + i\odot\beta_n(\theta + i\mathrm{H})^{[n]}]$$

where α_n , β_n are real constants. $\boldsymbol{R}_2$ can obviously be applied to finite sums and in some cases to infinite ones, producing potentials and stream functions of a compressible fluid flow.

REMARK 3.2. In operating with these symbols, the following rules hold

$$i[(\theta + i\mathrm{H})^{[n]}] = i\odot(\theta + i\mathrm{H})^{[n]}, \qquad i[i\odot(\theta + i\mathrm{H})^{[n]}] = -(\theta + i\mathrm{H})^{[n]}.$$

In a joint investigation, Bers and Gelbart [17] independently of the author of the present paper, found the same operator, which they then investigated in subsequent publications [18, 19]. They term the functions obtained $\sum$-monogenic.

4. *"Basic systems of solutions" and their transformation due to a circuit around the singularity of the coefficient N of equation* (1.5). In considering ordinary differ-

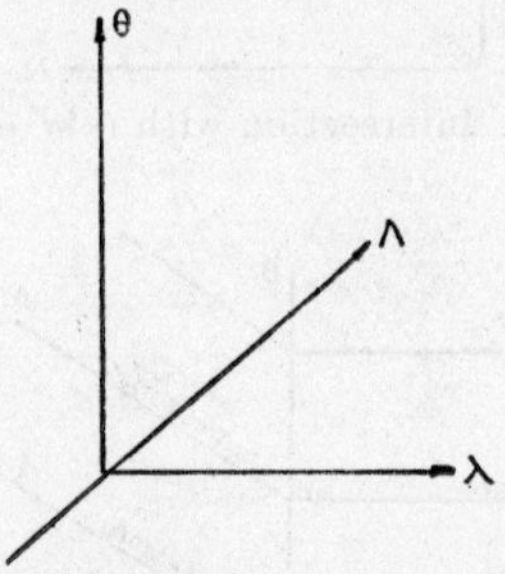

Fig. 3c. Intersection with $\vartheta = \vartheta_1 > 0$

ential equations whose coefficients have singularities of Fuchs type, we associate with each singularity a certain linear transformation. It seems of considerable interest to attempt to generalize these considerations to the case of partial differential equations of the type (1.5) [or (1.7)] where $N = N(\lambda)$ is a single-valued [or finitely many-valued] function which has a singularity on the line $\lambda = 0$. In these investigations, it is useful to continue the functions to complex values of the independent variables, that is, to consider $\psi^*(\lambda + i\Lambda, \theta + i\vartheta)$. See Remark 1.2, p. 20. The variables $\mathrm{Z} = (\lambda - \vartheta) + i(\Lambda + \theta)$, $\bar{\mathrm{Z}} = (\lambda + \vartheta) + i(\Lambda - \theta)$ become now two *independent* variables; they are conjugate to each other in the real plane, $\Lambda = 0$, $\vartheta = 0$. The coefficient F has as its only singularity the plane $\lambda = 0$, $\Lambda = 0$, on which a solution ψ^* can be singular. ψ^* may possess some singularities outside this surface. As can be shown, every singularity of ψ^*, except that in $\lambda = 0$, $\Lambda = 0$ must be a domain in a plane $\mathrm{Z} =$ const. (that is, in $[\lambda - \vartheta =$ const., $\Lambda + \theta =$ const.$]$) or in a plane $\bar{\mathrm{Z}} =$ const.; the boundary points of such a singularity domain lie either at infinity or belong to $[\lambda = 0, \Lambda = 0]$.

We shall now associate with the singularity plane of F a certain infinite matrix.

Let $P_0(Z_0, \bar{Z}_0)$ be a point which does not belong to $[\lambda = 0, \Lambda = 0]$ and let $\{\chi_\nu\}$ denote the system of functions which are obtained by orthogonalizing the system

$$\text{(3.24)} \qquad P[(Z - Z_0)^\nu], \qquad P[(\bar{Z} - \bar{Z}_0)^\nu], \qquad \nu = 0, 1, 2, \cdots,$$

$$P[(Z - Z_0)^\nu] = H\left[(Z - Z_0)^\nu - \int_{Z_0}^{Z}\int_{\bar{Z}_0}^{\bar{Z}} F Z_1^\nu \, dZ_1 \, d\bar{Z}_1 + \int_{Z_0}^{Z}\int_{\bar{Z}_0}^{\bar{Z}} \left(F \int_{Z_0}^{Z_1}\int_{\bar{Z}_0}^{\bar{Z}_1} F Z_2^\nu \, dZ_2 \, d\bar{Z}_2\right) dZ_1 \, d\bar{Z}_1 \cdots \right],$$

with respect to a hypersphere $\mathcal{H}^4$, whose center is at P_0 and whose radius is sufficiently small. As it is possible to show this system is complete for the class of solutions of (1.7) which are regular in $\mathcal{H}^4$. Suppose now $\{\chi_\nu^*\}$ represents another complete orthonormal system, then it is possible to develop every function χ_ν^* in the series $\chi_\nu^* = \sum_{\mu=1}^{\infty} \alpha_{\nu\mu}\chi_\mu$ (and $\chi_\nu = \sum_{\mu=1}^{\infty} \gamma_{\nu\mu}\chi_\mu^*$), so that we can associate with both systems an infinite matrix $\mathfrak{A} = \{\alpha_{\nu\mu}\}$. In particular, let us suppose that we have a closed curve $\mathcal{C}$ passing through P_0. Each $P[(Z - Z_0)^\nu]$ and $P[(\bar{Z} - \bar{Z}_0)^\nu]$ is defined for $|Z| < \infty$, $|\bar{Z}| < \infty$ except $[\lambda = 0, \Lambda = 0]$. Each of these functions is single-valued except possibly at $[\lambda = 0, \Lambda = 0]$ where it can have a branch point. Therefore if we start at P_0 with $\chi_\nu(Z, \bar{Z})$ and we move along a simple closed curve, which cannot be reduced to a point without intersecting $[\lambda = 0, \Lambda = 0]$ then going once around the curve we obtain, in general, new functions, say $\chi_\nu^{(1)}$, which form again a complete system. Each function $\chi_\nu^{(1)}$ can be developed in $\mathcal{H}^4$ into a series, $\chi_\nu^{(1)} = \sum_{\mu=1}^{\infty} \beta_{\nu\mu}\chi_\mu$, where $\mathfrak{B} = \{\beta_{\nu\mu}\}$ represents an infinite matrix. If instead of the χ_ν's we use some other complete system $\{\tau_\nu\}$ possessing the same properties as $\{\chi_\nu\}$, and if $\tau_\nu = \sum_{\mu=1}^{\infty} \gamma_{\nu\mu}\chi_\mu$, then we obtain that the matrix $\mathfrak{E} = \{\epsilon_{\nu\mu}\}$ corresponding to $\mathfrak{B}$ is similar to $\mathfrak{B}$, that is, we have $\mathfrak{E} = \mathfrak{D}^{-1}\mathfrak{B}\mathfrak{D}$, where $\mathfrak{D} = \{\gamma_{\nu\mu}\}$. It seems of interest to consider the solutions in the Riemann space of $F(\lambda)$ and since $F = O(\lambda^{-2})$ for $\lambda \to 0$, it is suitable to use functions $(Z - Z_0)^\nu$, $(\bar{Z} - \bar{Z}_0)^\nu$ as the system χ_ν.

Continuing this method of attack we can associate with the singularity surface $[\lambda = 0, \Lambda = 0]$ a certain infinite matrix. The study of its properties can be utilized for the study of the behavior of solutions of the compressibility equation. In some later place we shall return to this question.

4. The method of orthogonal functions and the theory of integral operators. As we indicated in §1, many problems in the theory of compressible fluids may be solved by the hodograph method which enables us to consider a linear equation instead of a non-linear one. In particular, as we indicated in §3, this method enables us to handle the mixed case. On the other hand, there are some problems in the two-dimensional case where it seems to be preferable to work directly in the physical plane. Further, in many instances, for example, in the case of three-dimensional motion, we are unable to reduce the equation to a linear one. Therefore, it is of importance to develop methods of working directly in the physical plane, in which case we have to solve a non-linear partial differential equation.

Recently, the author of the present paper (together with M. Schiffer) developed certain considerations which enable us to use methods of the theory of complex orthogonal functions for the solution of linear and non-linear partial differential equations. See [10, 14–16]. Naturally it is impossible, in the case of a non-linear equation, to introduce a linear operator which transforms in biuniform manner analytic functions of one complex variable into solutions of this equation and vice versa. But it is possible to define a larger class of functions which form a linear space such that the solutions of non-linear equations can be mapped into a certain subset of vectors in this linear space. This fact will enable us to derive certain properties of solutions of non-linear partial differential equations, in particular, those which appear in the theory of compressible fluids.

Suppose at first that the density ρ is known at every point of a flow past a profile $\mathfrak{P}$. Then the potential function[12] ϕ can be determined as the solution of the differential equation

$$(4.1)\qquad \frac{\partial}{\partial x}\left(\rho\,\frac{\partial\phi}{\partial x}\right)+\frac{\partial}{\partial y}\left(\rho\,\frac{\partial\phi}{\partial y}\right)\equiv\rho\left[\Delta\phi+\frac{\rho_x}{\rho}\frac{\partial\phi}{\partial x}+\frac{\rho_y}{\rho}\frac{\partial\phi}{\partial y}\right]=0$$

which is defined in $\mathfrak{B}$ (the exterior of $\mathfrak{P}$), satisfies on the boundary $\mathfrak{C}$ of $\mathfrak{B}$ the conditions

$$(4.2)\qquad (\partial\phi/\partial n)=0,$$

and at infinity possesses a singularity representing a doublet. A solution with such a singularity can be determined by methods similar to those of §2. Instead of ϕ we can introduce the function $\phi^*=\phi/h$, $h=\rho^{-1/2}$, ϕ^* satisfies in $\mathfrak{B}$ the equation

$$(4.3)\qquad L^*(\phi^*)\equiv\Delta\phi^*+4P\phi^*=0,\qquad P=-4^{-1}\rho^{-1/2}\Delta(\rho^{-1/2}),$$

satisfies the boundary condition

$$(4.4)\qquad B(\phi^*)\equiv(\partial\phi^*/\partial n)-q\phi^*=0,\qquad q=-h^{-1}(\partial h/\partial n),$$

and has a singularity at infinity. Let us assume that $P<0$. The determination of ϕ^* can be reduced to the determination of a function φ^* which satisfies (4.3) in $\mathfrak{B}$, is regular at infinity, and satisfies the condition $B(\varphi^*)=p$ on the boundary $\mathfrak{C}$. This can be achieved by the use of the theory of orthogonal functions. Let $a(x,y)$ and $b(x,y)$ be two functions twice differentiable in $\mathfrak{B}+\mathfrak{C}$, and such that

$$(4.5)\qquad \iint_{\mathfrak{B}}\left[\frac{\partial a}{\partial x}+\frac{\partial b}{\partial y}\right]dx\,dy=\int_{\mathfrak{C}}q\,ds,$$

$$a\cos(n,x)+b\cos(n,y)=q(s)\text{ on }\mathfrak{C}.$$

[12] We consider here the potential function rather than stream function, since in the three-dimensional case only the former can be defined.

We introduce now a complete system $\{\varphi_\nu^*\}$, $\nu = 1, 2,$ of solutions of (4.3) which are orthogonalized so that

$$R(\varphi_\nu^*, \varphi_\mu^*) \equiv \int\int_{\mathfrak{B}} \left[\frac{\partial \varphi_\nu^*}{\partial x} \frac{\partial \varphi_\mu^*}{\partial x} + \frac{\partial \varphi_\nu^*}{\partial y} \frac{\partial \varphi_\mu^*}{\partial y} - 4P \varphi_\nu^* \varphi_\mu^* \right. \tag{4.6}$$
$$\left. - \frac{\partial(a\varphi_\nu^* \varphi_\mu^*)}{\partial x} - \frac{\partial(b\varphi_\nu^* \varphi_\mu^*)}{\partial y} \right] dx\, dy = \delta_{\nu\mu} = 1 \text{ for } \nu = \mu, = 0 \text{ for } \nu \neq \mu.$$

Then every solution φ^* of (4.3) can be written in $\mathfrak{B}$ in the form

$$\varphi^* = \sum_{\nu=1}^{\infty} a_\nu \varphi_\nu^*, \qquad a_\nu = \int_{\mathfrak{C}} \left[\varphi_\nu^* \left(\frac{\partial \varphi^*}{\partial n} - q\varphi^* \right) \right] ds. \tag{4.7}$$

See [16, §6]. This result enables us to determine the desired potential function ϕ.

In order to solve the non-linear equation, we apply the method of successive approximations in the following manner. At first we determine as a first approximation, the potential function $\phi^{(1)}$ for the incompressible fluid case. By using then the formula

$$\rho^{(2)} = \left[1 - \frac{1}{2} \frac{(k-1)}{a^2} (\phi_x^{(1)2} + \phi_y^{(1)2}) \right]^{1/(k-1)} \tag{4.8}$$

we determine the second approximation for the density. Using $\rho^{(2)}(x, y)$ we determine the second approximation for ϕ, say $\phi^{(2)}$. Substituting it in the right-hand side of (4.8) we obtain the third approximation $\rho^{(3)}$ for ρ, and so on. As it can be shown the procedure converges in certain cases, in which it yields the desired solution of the non-linear differential equation.

Bibliography

1. Stefan Bergman, *Zur Theorie der Funktionen, die eine lineare partielle Differential gleichung befriedigen*. I. Rec. Math. (Mat. Sbornik) N. S. vol. 2 (1937) pp. 1169–1198.

2. ———, *The hodograph method in the theory of compressible fluids*, Supplement to *Fluid dynamics* by von Mises and Friedrichs, Brown University, 1942.

3. ———, *A formula for the stream function of certain flows*, Proc. Nat. Acad. Sci. U. S. A. vol. 29 (1943) pp. 276–281.

4. ———, *On two-dimensional flows of compressible fluids*, NACA Technical Note No. 972, 1945.

5. ———, *Graphical and analytical methods for the determination of a flow of a compressible fluid around an obstacle*, NACA Technical Note No. **973**, 1945.

6. ———, *Methods for the determination and computation of flow patterns of a compressible fluid*, NACA Technical Note No. 1018, 1946.

7. ———, *On supersonic and partially supersonic flows*, NACA Technical Note No. 1096, 1946.

8. ———, *Two-dimensional subsonic flows of a compressible fluid and their singularities*, Trans. Amer. Math. Soc. vol. 62 (1947) pp. 452–498.

9. ———, *Two-dimensional transonic flow patterns*, Amer. J. Math. vol. 70 (1948) pp. 856–891.

10. ———, *Functions satisfying certain partial differential equations of elliptic type and their representation*, Duke Math. J. vol. 14 (1947) pp. 349–366.

11. ———, *On tables for the determination of transonic flow patterns*, Hans Reissner Anniversary Volume, Ann Arbor, Edwards, 1949, pp. 13–36.

12. Stefan Bergman and B. Epstein, *Determination of a compressible fluid flow past an oval-shaped obstacle*, Journal of Mathematics and Physics vol. 26 (1948) pp. 195–222.

13. Stefan Bergman and L. Greenstone, *Numerical determination by use of special computational devices of an integral operator in the theory of compressible fluids.* I. *Determination of the coefficients of the integral operator by the use of punch card machines*, Journal of Mathematics and Physics vol. 26 (1947) pp. 1–9.

14. Stefan Bergman and M. Schiffer, *A representation of Green's and Neumann's functions in the theory of partial differential equations of second order*, Duke Math. J. vol. 14 (1947) pp. 609–638.

15. ———, *On Green's and Neumann's functions in the theory of partial differential equations*, Bull. Amer. Math. Soc. vol. 53 (1947) pp. 1141–1151.

16. ———, *Kernelfunctions in the theory of partial differential equations of elliptic type*, Duke Math. J. vol. 15 (1948) pp. 535–566.

17. L. Bers and A. Gelbart, *On a class of differential equations in mechanics of continua*, Quarterly of Applied Mathematics vol. 1 (1943) pp. 168–188.

18. ———, *On a class of functions defined by partial differential equations*, Trans. Amer. Math. Soc. vol. 56 (1944) pp. 67–93.

19. ———, *On generalized Laplace transformations*, Ann. of Math. vol. 48 (1947) pp. 342–357.

20. R. v. Mises and M. Schiffer, *On Bergman's integration method in two dimensional compressible flow*, Advances in Applied Mechanics vol. 1 (1948) pp. 249–285.

21. H. S. Tsien and Y. H. Kuo, *Two-dimensional irrotational mixed subsonic and supersonic flow of a compressible fluid and the upper critical Mach number*, NACA Technical Note No. 995, 1946.

HARVARD UNIVERSITY,
CAMBRIDGE, MASS.

AN EXISTENCE THEOREM IN TWO-DIMENSIONAL GAS DYNAMICS

BY

LIPMAN BERS

Introduction. The basic existence and uniqueness theorems of gas dynamics are yet to be established. Consider, for instance, a profile P in the x, y-plane. Does there exist a two-dimensional steady potential gas flow around P, such that the Kutta-Joukowsky condition is satisfied at the sharp trailing edge of P, the flow is uniform at infinity and possesses there a given direction, and the speed q attains somewhere along P the preassigned maximum $q_{\max}$? The answer to this question is not known, though it is known that it cannot be always in the affirmative for supersonic values of $q_{\max}$.[1] It is generally believed that a uniquely determined flow exists whenever $q_{\max}$ is less than the critical speed. The only known result in this direction, however, is that due to Frankl and Keldysh [6]. It asserts the existence and uniqueness of a flow for every preassigned *sufficiently small* value of the free stream speed q_∞. A more special theorem of Slioskin [13], proved in the framework of Chaplygin's approximation and for symmetrical flows only, is of the same general character.[2] We want to show how Chaplygin's approximation leads to a complete existence theorem for convex profiles.

Chaplygin's approximation. The density of a potential adiabatic flow is given by[3]

$$\rho = \left(1 - \frac{\gamma - 1}{2} q^2\right)^{1/(\gamma-1)} \tag{1}$$

where γ is the ratio of the specific heats. Chaplygin's approximation method [4] consists of setting $\gamma = -1$, so that (1) becomes

$$\rho = (1 + q^2)^{-1/2}. \tag{2}$$

It is well known how this seemingly paradoxical assumption can be justified, and that in some cases it leads to results which are of practical value [9]. We state some of the known consequences of assumption (2). The continuity equation for the velocity potential φ becomes the equation of a minimal surface:

$$(1 + \varphi_y^2)\varphi_{xx} - 2\varphi_x\varphi_y\varphi_{xy} + (1 + \varphi_x^2)\varphi_{yy} = 0.$$

The complex potential $F = \varphi + i\psi$ (ψ being the stream function) becomes an analytic function of the "distorted complex velocity"

[1] This follows from a recent non-existence theorem of Nikolsky and Taganov [12].

[2] Slioskin also considers a free boundary problem. His treatment is based on an integro-differential equation different from the one used by us.

[3] We assume that the units are chosen so that the stagnation density and the speed of sound at a stagnation point are both equal to unity.

$$w^* = q^* e^{-i\theta} \tag{3}$$

where $\theta = \arctan(\varphi_y/\varphi_x)$ is the inclination of the velocity vector, and the "distorted speed" q^* is defined by

$$q^* = \frac{q}{1 + (1 + q^2)^{1/2}}. \tag{3'}$$

The complex variables F, w^* and $z = x + iy$ are connected by the relation

$$dz = \frac{1}{2}\left(\frac{dF}{w^*} - \overline{w^*\, dF}\right). \tag{4}$$

This relation is the key to the "indirect" problem of airfoil theory (construction of flows around profiles whose shapes are determined afterwards). Tsien [15] solves this problem by taking for F the complex potential of an incompressible flow past some profile in an auxiliary ζ-plane, and for w^* its complex velocity: $w^* = dF/d\zeta$. This procedure fails for circulatory flows. The author showed [1] that circulatory flows can be obtained by setting $w^* = (dF/d\zeta)^n$, n being an appropriately chosen constant. A more elegant solution of the indirect problem is due to Gelbart [7] who takes for w^* an essentially arbitrary function of ζ subject only to a few conditions.[4]

The direct problem. We are concerned here with the direct problem where the profile P is preassigned.[5] In the case of an incompressible fluid this problem is solved by mapping the domain $E(P)$ exterior to the profile P onto the domain $|\zeta| > 1$, by means of a function which is analytic in z, and hence also in F. The same method may be applied to a Chaplygin flow. We assume the existence of a solution and map $E(P)$ onto $|\zeta| > 1$ by a function which is analytic in F (but, of course, not in z). The existence of such a mapping (subject to the usual normalization) is easily established. In the ζ-plane both F and w^* are analytic functions.

To simplify the argument we consider the most primitive case: a symmetrical flow past a symmetrical profile without corners. We denote the arc-length on P by σ, and assume (without loss of generality) that the length of P is 2π. The mapping of P onto the circle $\zeta = e^{i\omega}$ is given by a function

$$\sigma = f(\omega), \qquad 0 \leqq \omega \leqq 2\pi. \tag{5}$$

Assume that we know this function. Since along P the slope of the velocity vector coincides with that of the profile we know the boundary values of the harmonic function $\theta(\zeta)$ along the circle $|\zeta| = 1$. From these values we may compute the boundary values of the conjugate harmonic function $\log q^*$, and once we know q^* along the circle we also know the velocity distribution q along P.

[4] Gelbart's solution was later obtained independently by Lin [10] and Germain [8]. A solution of the indirect problem was also given by Christianovich and Yuriev [5].

[5] From an abstract presented to the American Mathematical Society [14] it is seen that the same or similar problem has been attacked by Tschen by methods of the calculus of variations.

The same velocity distribution can be computed in a different way. It is easy to verify that in the ζ-plane F satisfies the boundary conditions for a complex potential of a flow around a circle. Thus

$$F = A(\zeta + 1/\zeta) \tag{6}$$

where A is a positive constant. On P

$$q = \left| \frac{d\varphi}{d\sigma} \right| = \left| \frac{d\varphi}{d\omega} \frac{d\omega}{d\sigma} \right|$$

where $d\varphi/d\omega$ is to be evaluated along $|\zeta| = 1$. From (5) and (6) it follows that

$$q = \frac{2A \mid \sin \omega \mid}{f'(\omega)}.$$

The function (5) must be such that the two determinations of q lead to the same values. If we write down this condition, we obtain an integral equation for the function $f(\omega)$. We omit all intermediate computations (which may be found in [2]) and state only the final result.

Let the equation of the profile P be written in the form

$$z = Z(\sigma), \qquad 0 \leqq \sigma \leqq 2\pi,$$

and let the function $\Theta(\sigma)$ be defined by

$$\exp [i\Theta(\sigma)] = Z'(\sigma).$$

Associate to each function $f(\omega)$ the function

$$h(\omega) = \frac{1}{2\pi} \int_0^\pi \{\Theta[f(\omega + t)] - \Theta[f(\omega - t)] - 2t\} \cot \frac{t}{2}\, dt \tag{7}$$

and the number

$$\lambda = \frac{q_{\max}/[1 + (1 + q_{\max}^2)^{1/2}]}{\max \mid 2 \sin \omega \cdot e^{h(\omega)} \mid} \tag{8}$$

where $q_{\max}$ is the prescribed value of the maximum speed. The integral equation for $f(\omega)$ may be written in the form

$$f(\omega) = \int_0^\omega H(\omega')\, d\omega' \Big/ \frac{1}{2\pi} \int_0^{2\pi} H(\omega')\, d\omega' \tag{9}$$

where

$$H(\omega) = e^{-h(\omega)} - 4\lambda^2 \sin^2 \omega \cdot e^{h(\omega)}. \tag{10}$$

It might be mentioned that this equation can be solved numerically, though the convergence of the iteration procedure used has not yet been established.[6]

[6] The procedure is described in [2]. An iteration method for circulatory flows will be described in a forthcoming paper. An approximate numerical method for solving the direct problem is contained in a forthcoming paper by Gelbart and Resch.

Application of the Schauder-Leray method. In order to apply to our equation the topological method of Schauder and Leray we transform it into an equation for the derivative $g(\omega) = f'(\omega)$. Let $\boldsymbol{S}(f)$ denote the right-hand side of (7) and $\boldsymbol{R}(h)$ the right-hand side of (10). The operator $\boldsymbol{S}$ depends on the function $\Theta(\sigma)$, that is, on the shape of P. The operator $\boldsymbol{R}$ depends on the value of $q_{\max}$ since it involves the parameter λ determined by (8). We also define the operator $\tau = \boldsymbol{T}(t)$ by the equation

$$\tau(\omega) = t(\omega) \Big/ \frac{1}{2\pi} \int_0^{2\pi} |\, t(\omega') \,|\, d\omega'.$$

Now set

$$\boldsymbol{F} = \boldsymbol{TRST}.$$

Equation (9) is clearly equivalent to the following equation for the derivative g:

$$g = \boldsymbol{F}(g). \tag{11}$$

We introduce in (11) a real parameter k, $0 \leqq k \leqq 1$, by setting

$$\boldsymbol{F}(k, g) = 2\pi(1 - k) + k\boldsymbol{F}(g),$$

and forming the equation

$$g - \boldsymbol{F}(k, g) = 0. \tag{12}$$

The operator $\boldsymbol{F}(k, g)$ is defined whenever $g(\omega)$ is a (not identically vanishing) continuous function defined on the closed interval $(0, 2\pi)$.

Consider the Banach space of continuous functions $g(\omega)$, $0 \leqq \omega \leqq 2\pi$, with the norm $\| g \| = \max |\, g(\omega) \,|$. Let δ be a sufficiently small positive number, and let $\bar{\Omega}_\delta$ denote the closed subdomain of the Banach space consisting of all functions satisfying the conditions

$$\| g \| \leqq \frac{8}{\delta}, \qquad \int_0^{2\pi} |\, g(\omega) \,|\, d\omega \geqq \delta.$$

Under the assumption that P is convex (that is, that $\Theta' \geqq 0$) and with the aid of some results from the theory of conjugate functions it can be shown that $\boldsymbol{F}(k, g)$ is continuous in k and g for $0 \leqq k \leqq 1$ and g in $\bar{\Omega}_\delta$, and completely continuous in g. It also can be shown that for no k between 0 and 1 does the equation (12) possess a solution on the boundary of $\bar{\Omega}_\delta$. Since for $k = 0$ the operator $\boldsymbol{F}(k, g)$ is a simple translation and since the function $g \equiv 2\pi$ is an element $\bar{\Omega}_\delta$, we may apply the fundamental theorem of Schauder-Leray [11, in particular p. 63] to conclude that equation (12) possesses a solution for $k = 1$. But for this value of k equation (12) is identical with (11).

The existence theorem. If the profile is asymmetric and possesses a sharp trailing edge, and if the angle of attack is different from zero, a similar procedure may be used. Furthermore, it can be shown that if the integral equation has a solution, so does the original boundary value problem. The net result is the following existence theorem.

Let there be given a profile P and a point z_T on P. Let P be convex and continuously curved, except, perhaps, for a sharp trailing edge at z_T. For every angle of attack and every value of the maximum local speed there exists a potential gas flow around P which is uniform at infinity, satisfies the Chaplygin density speed relation, and possesses at z_T a stagnation point.

If P has a cusp at z_T, the theorem remains true, provided that the condition that z_T be a stagnation point is replaced by the condition that at z_T a streamline should divide itself into two branches.

The same method also yields other existence theorems. A detailed presentation will be published at a later date.

Non-existence of a flow past a corner. A few words should be said concerning the conditions imposed on the profile. The convexity condition is probably superfluous and due to an imperfection in the proof. The condition that the profile should be continuously curved could be relaxed easily. On the other hand, the condition that the profile possesses at most one sharp corner is essential.[7]

In fact, a Chaplygin flow past a convex corner is impossible. This statement is not a consequence of physical considerations, but a mathematical theorem expressing a property of minimal surfaces. The flow is impossible even if the speed is permitted to become infinite at the vertex of the corner.

To verify this theorem, assume that a Chaplygin flow past a solid wall W containing a convex corner at the point V is given. We assume that W has a continuously turning tangent, except at the point V where the tangent experiences a jump, and that in the neighborhood of V the streamlines are topologically equivalent to the lines $\operatorname{Im} z^{\alpha} = 0, 0 < \alpha < 1$. Let F be the complex potential of our flow and let $w^* = q^* e^{-i\theta}$ be its distorted complex velocity.[8] Consider a domain D bounded by an arc of a small circle with center at V and by an arc of W. The mapping $\zeta = F(z)$ takes D into a domain D' bounded by a continuous arc A' (image of the circular arc) and by a straight segment W' (image of the arc of W). The arc W' contains the image V' of V. Consider w^* as a function of ζ. It is analytic in D' and continuous at all points of the boundary $A' + W'$, except for the point V'. Now we map D' conformally onto the domain $|Z| < 1$, by means of an analytic function $Z = Z(\zeta)$ such that $Z(V') = 1$. In $|Z| < 1$ the function w^* is analytic. It is continuous for $|Z| = 1, Z \neq 1$. The argument $-\theta$ is continuous for $Z = e^{it}, t \neq 2n\pi$. The limits $\theta(e^{+i0})$ and $\theta(e^{-i0})$ exist and $\theta(e^{+i0}) - \theta(e^{-i0}) > 0$. It follows that w^* is singular at $Z = 1$ and $|w^*| = q^*$ becomes infinite at this point. Going back to the domain D we see that q^* attains arbitrarily large values in the neighborhood of V. But this is impossible since, by virtue of (3′), q^* cannot exceed 1.

Open questions. Besides being restricted to convex profiles our result is

[7] Our proof can be easily extended to the case of two convex corners, except that there the angle of attack can not be prescribed. It must be chosen so that both corners become stagnation points.

[8] We require that F be continuous on W, ψ monotonic on every stream line and w^* continuous on W except at V.

incomplete in that it contains no statement concerning the uniqueness of the solution. It seems probable that the solution *is* unique. It also would be of interest to show that the value of q_∞ may be prescribed instead of that of q_{max}.

Of greater importance is the question whether the existence proof for subsonic flows governed by the "exact" adiabatic density-speed relation (1) can be obtained by similar methods. The first step would consist in reducing the boundary value problem to a mapping problem and to a functional equation for a function of a single real variable. For symmetrical flows this has been accomplished [3].

(Note added in proof.) I am now able to replace the condition that P be convex by a less restrictive condition which is satisfied by aerodynamically significant shapes.

Bibliography

1. L. Bers, NACA TN No. 969, 1945.
2. ———, NACA TN No. 1006, 1946.
3. ———, NACA TN No. 1012, 1946.
4. S. A. Chaplygin, Scientific Memoirs of the Imperior University at Moscow, Section of Mathematics and Physics vol. 21 (1902) pp. 1–21. Also NACA TM No. 1063, 1944.
5. S. A. Christianovich and I. M. Yuriev, Applied Mathematics and Mechanics vol. 11 (1947) pp. 105–118.
6. E. I. Frankl and M. Keldysh, Bull. Acad. Sci. URSS. vol. 12 (1934) pp. 561–601.
7. A. Gelbart, NACA TN No. 1170, 1947.
8. P. Germain, C. R. Acad. Sci. Paris vol. 223 (1947) pp. 532–534.
9. Th. von Kármán, Journal for the Aeronautical Sciences vol. 8 (1941) pp. 337–356.
10. C. C. Lin, Quarterly of Applied Mathematics vol. 4 (1946) pp. 291–297.
11. T. Leray and T. Schauder, Ann. École Norm vol. 51 (1934) pp. 46–78.
12. A. A. Nikolski and G. I. Taganov, Applied Mathematics and Mechanics vol. 10 (1946) pp. 482–502.
13. N. A. Slioskin, Scientific Memoirs of the Moscow State University vol. 7 (1937) pp. 43–69.
14. Y. W. Tschen, Bull. Amer. Math. Soc. vol. 53 (1947) p. 280.
15. H. S. Tsien, Journal for the Aeronautical Sciences vol. 6 (1939) pp. 339–407.

Syracuse University,
Syracuse, N. Y.

RECENT DEVELOPMENTS IN FREE BOUNDARY THEORY

BY

GARRETT BIRKHOFF

It has been known for some time by practical men that "wake theory" does not describe actual physical wakes. Recent intensive study of "cavity motion" supports the belief that it may however describe the high-speed motion of a rigid body through water, when an air- or vapor-filled *cavity* of very low density follows in the "wake" of the body.

In collaboration with Lynn Loomis and Milton Plesset, the author has shown that if this is so, very large wall corrections must be made in water tunnel measurements of cavity drag coefficients. It is noteworthy that these corrections become negligible if $C_D = 2D/\rho v^2$ is computed using the velocity v along the free boundary, instead of the usual mean upstream velocity. Gilbarg and Rock have shown that earlier results of Riabouchinsky and a new "reentrant jet" model both predict a variation $C_D(K)$ in drag coefficient C_D with the "cavitation parameter" K, according to the formula $C_D(K) = (1 + K)C_D(0)$; this prediction is confirmed roughly by experiment. The author and T. E. Caywood have constructed a simple mathematical model for jets formed from a collapsing cavity. By a simple transformation, this gives a model of a new type for the penetration of a liquid of one density by a jet of a different density.

However, experiments of R. M. Davies and L. B. Slichter show that cavity viscosity is not always negligible; moreover a ratio ρ'/ρ of cavity density to liquid density as low as .001 may exert a strong influence on cavity behavior.

Furthermore, the mathematical results described earlier cover only *two-dimensional* motion. The achievements of N. Levinson in determining the asymptotic shape of an axially symmetric cavity, and of R. V. Southwell in computing by "relaxation methods" specific axially symmetric "wake" profiles (jets from a circular orifice had been treated by Trefftz), should be noted.

A full exposition of the preceding material will appear in the author's forthcoming Taft Lectures on *Topics in Fluid Dynamics*, to be published by the Princeton University Press. These Lectures will also include a description of the application of free boundary theory to lined hollow charges such as the "Bazooka", by Sir Geoffrey Taylor, D. P. MacDougall, E. M. Pugh, and the author. This material has been cleared for publication since the Brown Symposium.

HARVARD UNIVERSITY,
CAMBRIDGE, MASS.

THEORY OF THE PROPAGATION OF SHOCK WAVES FROM CYLINDRICAL CHARGES OF EXPLOSIVE

BY

STUART R. BRINKLEY, JR. AND JOHN G. KIRKWOOD

We describe an approximate theory of the propagation of shock waves from infinitely long cylinders of explosive which is valid in any exterior medium. The theory takes account of the finite entropy increment in the fluid resulting from the passage of the shock wave and it permits the use of the exact Hugoniot curves for the fluid. Two cases are considered: (a) The one-dimensional theory resulting from the assumption of adiabatic isometric conversion of the entire explosive charge to its decomposition products, for which the shock wave properties are functions of time and the radial coordinate only, and (b) the two-dimensional theory resulting from a consideration of the shock wave produced by a stationary detonation wave traveling in the axial direction of the cylinder with finite velocity. In the latter case also, the shock wave properties are functions of time and the radial coordinate only, since the axial and radial coordinates are connected by a relation, $z = \zeta(r)$.

The basic equations. The Eulerian equations of motion and continuity for an inviscous fluid are

$$(1) \qquad \frac{1}{\rho c^2}\frac{Dp}{Dt} = -\nabla\cdot\boldsymbol{u}, \qquad \frac{D\boldsymbol{u}}{Dt} = -\frac{1}{\rho}\nabla p,$$

where $\boldsymbol{u}$ is the particle velocity, p the pressure in excess of the pressure p_0 of the undisturbed fluid, ρ the density, and c the Euler sound velocity. They are to be solved subject to initial conditions specified on a curve in the $\boldsymbol{r}$, t-space ($\boldsymbol{r}$ is the Euler position vector) and to the Rankine-Hugoniot[1] equations which constitute supernumerary boundary conditions at the shock front,

$$(2) \qquad p = \rho_0 uU, \qquad \rho_0/\rho = 1 - p/\rho_0 U^2, \qquad \Delta H = (p/2)(1/\rho_0 + 1/\rho),$$

where ΔH is the specific enthalpy increment experienced by the fluid in traversing the shock front, U the velocity of the shock front in the direction of its normal, u the component of particle velocity normal to the shock front, and ρ_0 the density of the undisturbed fluid. Equations (2) are compatible with equations (1) and the specified initial conditions only if the shock front follows an implicitly prescribed curve $\boldsymbol{R}(t)$ in the $\boldsymbol{r}$, t-space. Equations (1) and (2) are supplemented by the entropy transport equation $DS/Dt = 0$, which we shall not explicitly use, and by an equation of state of the fluid $p = p(H, \rho)$ which, with equations (2), permits the evaluation of all the properties of the shock front as functions of the peak pressure p.

---

[1] W. J. M. Rankine, Transactions of the Royal Society of London, vol. A160 (1870) p. 277. H. Hugoniot, J. École Polytech. vol. 57 (1887) p. 3; vol. 58 (1888) p. 1.

For the system with axial symmetry, we denote an operator which follows the shock front by

$$\frac{d}{dR} = \left[\boldsymbol{l}_r \cdot \left(\nabla_0 \boldsymbol{r} \cdot \nabla + \frac{1}{U} \boldsymbol{n} \frac{D}{Dt} \right) \right]_{r=R}, \tag{3}$$

where R is the radial coordinate of the shock front, $\boldsymbol{l}_r$ and $\boldsymbol{n}$ are unit vectors in the radial direction and in the direction of the normal to the shock front, respectively, and $\nabla_0 \boldsymbol{r}$ is the deformation-rotation dyadic. The components of $\nabla_0 \boldsymbol{r}$ are easily found at the shock front from the fact that the medium experiences a pure strain of magnitude $\rho_0/\rho - 1$ in a direction normal to the shock front as the result of the passage of the wave. If the operator d/dR is applied to the first of equations (2) and to the relation describing continuity of the tangential component of particle velocity at the shock front, there result

$$g \frac{dp}{dR} = \rho_0 U \left(\boldsymbol{u} \cdot \frac{d\boldsymbol{n}}{dR} + \frac{d\boldsymbol{u}}{dR} \cdot \boldsymbol{n} \right),$$

$$\frac{d\boldsymbol{u}}{dR} \cdot (\boldsymbol{l}_\varphi \times \boldsymbol{n}) = \boldsymbol{u} \cdot \left(\frac{d\boldsymbol{n}}{dR} \times \boldsymbol{l}_\varphi \right), \tag{4}$$

$$g = g(p) = 1 - \frac{p}{U} \frac{dU}{dp}.$$

In the appendix, it is shown that the conservation of energy may be expressed by

$$\left(\frac{Dp}{Dt} \right)_{r=R} + \rho_0 U \boldsymbol{n} \cdot \left\{ \frac{D\boldsymbol{u}}{Dt} + \boldsymbol{l}_\varphi \times [(\boldsymbol{l}_z \cdot \nabla_0 \boldsymbol{r}) \cdot \nabla_0 \boldsymbol{u}] \right\}_{r=R} = -u \sin \vartheta \left[\frac{\nu R p^2}{K(R)} + \frac{p}{R} \right]_{r=R}, \tag{5}$$

where $\nabla_0 \boldsymbol{u}$ is the Euler rate of strain dyadic, ϑ is the angle between the tangent to the shock front and the r-axis, and where each term is evaluated at the shock front. The shock wave energy at R per unit area of initial generating surface is $K(R)/a_0$, where a_0 is the Lagrange radial coordinate of the generating surface and $K(R)$ is given by

$$K(R) = \int_R^\infty \rho_0 r_0 \, h[p(r_0)] \, dr_0, \tag{6}$$

with $h(p)$ the specific enthalpy increment of the fluid at pressure p_0 for the entropy increment corresponding to shock front pressure p. The quantity ν is the reduced Lagrange energy-time integral,

$$\nu(R) = \int_0^\infty f(R, \tau) \, d\tau, \qquad \tau = \frac{t - t_0(R)}{\mu}, \tag{7}$$

where $t_0(R)$ is the time of arrival of the shock front at the point $r = R$, $-1/\mu$ is the initial logarithmic slope of the Lagrange energy-time curve, and $f(R, \tau)$ is the energy-time integrand, normalized by its peak value at the shock front, expressed as a function of R and the reduced time τ which normalizes its initial slope to -1 if μ does not vanish. We assume that f is a monotone decreasing function of τ. Equation (7) is exact, involving integrals of equations (1) for the evaluation of f. However, if $f(a_0, \tau)$ is initially a monotone decreasing function of τ, $f(R, \tau)$ will remain so and will at large R become asymptotically a quadratic function of τ.[2] This implies that ν is a slowly varying function of R for which sufficiently accurate estimates for many purposes can be made without explicit integration of equations (1). The assignment of a value independent of R to ν is equivalent to imposing a similarity restraint on the energy-time curve.

The initial energy-time curve of an explosion wave is rapidly decreasing. An expansion of the logarithm of the function in a Taylor series in the time, the peak approximation, is appropriate for an initial estimate of ν. This corresponds to an exponential $f(\tau)$,

$$f(\tau) = e^{-\tau},$$

and results in $\nu = 1$. For the asymptotic quadratic energy-time curve,

$$f(\tau) = (1 - \tau/2)^2, \ \tau \leqq 2; \qquad f(\tau) = 0, \ \tau > 2,$$

which leads to the value $\nu = 2/3$. As a convenient empirical interpolation formula between the two extreme values, we have employed the relation

$$\nu = 1 - 3^{-1}\exp\,[-p/p_0]. \tag{8}$$

Propagation equations for the one-dimensional wave. The origin of the radial coordinate r is taken to be the axis of the generating cylinder. Equations (1), specialized to the shock front $r = R$, together with equations (4) and (5), provide four nonhomogeneous, linear relations between the four nonvanishing time and distance derivatives of pressure and particle velocity, evaluated at the shock front, with coefficients that can be expressed as functions of distance and peak pressure through equations (2) and the equation of state. The equations can be solved for the derivatives and an ordinary differential equation, $dp/dR = F(p, R)$, formulated with the aid of equation (3). An additional ordinary differential equation for the shock wave energy is obtained by differentiating equation (16). The results are

$$\frac{dp}{dR} = -\frac{\nu R p^3}{K(R)} M(p) - \frac{p}{2R} N(p), \tag{9}$$

$$\frac{dK}{dR} = -\rho_0 R h[p(R)], \tag{10}$$

[2] J. G. Kirkwood and H. A. Bethe (1941).

where

$$M(p) = \frac{1}{\rho_0 U^2} \frac{G}{2(1+g) - G},$$

(11) $$N(p) = \frac{4(\rho_0/\rho) + 2(1 - \rho_0/\rho)\,G}{2(1+g) - G},$$

$$G(p) = 1 - (\rho_0 U/\rho c)^2.$$

Equations (9) and (10) may be integrated numerically, employing tables of the functions $h(p)$, $M(p)$, $N(p)$, which can be constructed by numerical methods from the exact Hugoniot curves for the fluid. A detailed description of the one-dimensional theory, including plane and spherical waves, has recently been published.[3]

Propagation equations for the two-dimensional wave. We let r and z be the cylindrical coordinates relative to an origin in the detonation front with the z-axis coincident with the axis of the cylinder. The volocity of the detonation wave relative to a stationary origin in the negative z-direction is D. The profile of the shock wave is a surface of revolution $z = \zeta(R)$ with the differential equation

(12) $$\frac{d\zeta}{dR} = \tan \vartheta.$$

Since the distance traveled by the shock front in time dt in the direction of its normal is $U\,dt$, and in the same time, the origin of the coordinate system travels a distance $D\,dt$ in the negative z-direction,

(13) $$\cos \vartheta = U/D.$$

Taylor[4] has shown that the Chapman-Jouget[5] conditions can be satisfied at the front of a stationary detonation wave by solutions of the equations of hydrodynamics which depend only on $\mathbf{r}/t$. For solutions of the Taylor type,

(14) $$D/Dt = (\mathbf{u} - \mathbf{r}/t)\cdot\nabla.$$

The solutions of equations (1) and (14) for the exterior medium are compatible with the conditions on the boundary between explosion products and the exterior medium if solutions of the Taylor type are valid in the explosion products behind the detonation wave.

At any finite distance from an infinite cylinder of explosive, $\mathbf{r}/t = 0$, and

(15) $$D/Dt = \mathbf{u}\cdot\nabla.$$

Equation (15) can be employed to provide three relations between the deriva-

[3] S. R. Brinkley, Jr., and J. G. Kirkwood, Physical Reviews vol. 71 (1947) p. 606.

[4] G. I. Taylor, 1941.

[5] D. L. Chapman, Philosophical Magazine (5) vol. 47 (1889) p. 90. E. Jouget, C. R. Acad. Sci. Paris vol. 132 (1901) p. 573. For a discussion, see H. L. Dryden, F. D. Murnaghan, H. Bateman, Bulletin of the National Research Council, no. 84, 1932, p. 551.

tives with respect to time and the distance coordinates of the pressure and the components of the particle velocity. These relations and equations (1), specialized to the shock front, together with equations (4) and (5), provide nine nonhomogeneous linear equations between nine partial derivatives with coefficients that are functions of R, D, and p. The equations can be solved for the derivatives and an ordinary differential equation, $dp/dR = F(p, D, R)$, formulated with the aid of equation (3). The result is

$$\frac{dp}{dR} = \left\{ -\frac{\nu R p^3}{K(R)} M(p) - \frac{p}{2R} N(p) \right\} \Phi(p, D), \tag{16}$$

where

$$\Phi(p, D) = \left\{ 1 - \frac{\rho_0}{\rho} \frac{2(1-g)+G}{2(1-g)-G} \frac{U^2}{D^2 - U^2} \right\}^{-1}, \tag{17}$$

and where $K(R)$ is given by Equation (10) and $M(p)$, $N(p)$ by equations (11). When $p(R)$ is known, the profile $\zeta(R)$ of the shock front can be obtained by an auxiliary integration

$$\zeta(R) = \int_{a_0}^{R} \left\{ \frac{U[p(r_0)]^2}{D^2} - 1 \right\}^{1/2} dr_0 . \tag{18}$$

We note that $\text{Lim}_{D\to\infty} \Phi = 1$, and equation (16) is identical with equation (9) in this limit. Also, $\text{Lim}_{p\to 0} \Phi = 1$, and the asymptotic solutions of equations (9) and (16) have the same form.

A more extended discussion of the two-dimensional theory will appear elsewhere.

Appendix

Derivation of the energy equation. For an inviscous fluid, the adiabatic work w_0 per unit area of initial generating surface done on the fluid exterior to a generating cylinder is given by

$$\text{(a)} \quad 2\pi a_0 \, dz_0 \, w_0 = \int_{a_0}^{R} \rho_0 E[p(r_0)] 2\pi r_0 \, dz_0 \, dr_0 + (R, \zeta) \int_{t_0(R)}^{\infty} (p' + p_0) \mathbf{u}' \cdot d\mathbf{A}' \, dt,$$

where $\mathbf{u}'$ and p denote particle velocity and excess pressure behind the shock front (unprimed symbols being reserved for quantities at the shock front), $t_0(R)$ is the time of arrival of the shock front at the point with Lagrange cylindrical coordinates $r_0 = R$, $z_0 = \zeta(R)$, $E(p)$ is the specific energy increment of the fluid at pressure p_0 for the entropy increment corresponding to peak pressure p, a_0 is the radial Lagrange coordinate of the generating surface, and $d\mathbf{A}'$ is the Euler area element into which the Lagrange area element $2\pi r_0 \, dz_0 \mathbf{l}_r$ is transformed by the passage of the shock wave. Now

$$\text{(b)} \qquad (R, \zeta) \int_{t_0(R)}^{\infty} \mathbf{u}' \cdot d\mathbf{A}' \, dt = (R, \zeta) \int_{R}^{r(R,\zeta,\infty)} d\mathbf{r}' \cdot d\mathbf{A}',$$

where $\boldsymbol{r}'$ is the Euler position vector of the area element, $d\boldsymbol{A}'$, $r(r_0, z_0, t)$ is the Euler radial coordinate at time t, and where the variable of integration is restricted by the path and the definition of $d\boldsymbol{A}'$ to the radial coordinate. The integrand of equation (b) can be shown to be an exact differential of r_0 and t along any path of constant z_0. Accordingly, the path of integration can be changed to

$$[t_0(R), \zeta] \int_R^{a_0} + (a_0, \zeta) \int_{a_0}^{r(a_0,\zeta,\infty)} + (t, \zeta) \int_{r(a_0,\zeta,\infty)}^{r(R,\zeta,\infty)} .$$

Now,

$$[t_0(R), \zeta] \int_R^{a_0} d\boldsymbol{r}' \cdot d\boldsymbol{A}' = - \int_{a_0}^{R} 2\pi r_0 \, dz_0 \, dr_0 ,$$

since Euler and Lagrange coordinates are identical at or ahead of the shock front, and

$$(t, \zeta) \int_{r(a_0,\zeta,\infty)}^{r(R,\zeta,\infty)} d\boldsymbol{r}' \cdot d\boldsymbol{A}' = \int_{a_0}^{R} | \nabla_0 \boldsymbol{r} | \, 2\pi r_0 \, dz_0 \, dr_0 .$$

With the equation of continuity, $\rho_0 = | \nabla_0 \boldsymbol{r} | \rho$, equation (b) becomes

$$\text{(c)} \quad (R, \zeta) \int_{t_0(R)}^{\infty} \boldsymbol{u}' \cdot d\boldsymbol{A}' \, dt = (a_0, \zeta) \int_{t_0(a_0)}^{\infty} \boldsymbol{u}' \cdot d\boldsymbol{A}' \, dt + \int_{a_0}^{R} \left(\frac{\rho_0}{\rho} - 1 \right) 2\pi r_0 \, dz_0 \, dr_0 .$$

Combining equation (c) with equation (a) and introducing the dissipated enthalpy $h(p) = E(p) + p_0 \Delta(1/\rho)$, one obtains

$$\text{(d)} \quad \begin{aligned} &2\pi a_0 \, dz_0 \, w_0 - \int_{t_0(a_0)}^{\infty} p_0 \, \boldsymbol{u}' \cdot d\boldsymbol{A}' \, dt \\ &\qquad = \int_{a_0}^{R} \rho_0 h[p(r_0)] 2\pi r_0 \, dz_0 \, dr_0 + \int_{t_0(R)}^{\infty} p' \boldsymbol{u}' \cdot d\boldsymbol{A}' \, dt. \end{aligned}$$

The time integral of the right-hand member may be assumed to vanish for $R = \infty$. If one subtracts from equation (d) the expression obtained from that equation for $R = \infty$, there results the relation

$$\text{(e)} \qquad 2\pi dz_0 \, K(R) = \int_{t_0(R)}^{\infty} p' \boldsymbol{u}' \cdot d\boldsymbol{A}' \, dt,$$

where

$$\text{(f)} \qquad K(R) = \int_R^{\infty} \rho_0 h[p(r_0)] r_0 \, dr_0 .$$

Now, $d\boldsymbol{A}' = 2\pi r' ds' \boldsymbol{n}'$, where $\boldsymbol{n}'$ is the unit normal to ds' into which $dz_0 \boldsymbol{l}_z$ is transformed by the passage of the wave, and therefore,

$$d\boldsymbol{A}' = 2\pi r' dz_0 \boldsymbol{l}_\varphi \times (\boldsymbol{l}_z \cdot \nabla_0 \boldsymbol{r})'.$$

Equation (e) becomes[6]

$$\text{(g)} \qquad K(R) = \int_{t_0(R)}^{\infty} \boldsymbol{u}' \cdot [\boldsymbol{l}_\varphi \times (\boldsymbol{l}_z \cdot \nabla_0 \boldsymbol{r})'] r' p' \, dt.$$

The energy-time integral can be expressed in reduced form,

$$K(R) = F\mu\nu, \qquad 1/\mu = -(D/Dt \log F')_{t_0(R)},$$

$$\text{(h)} \qquad F = \boldsymbol{u} \cdot [\boldsymbol{l}_\varphi \times (\boldsymbol{l}_z \cdot \nabla_0 \boldsymbol{r})] \, rp, \; \tau = \frac{t - t_0(R)}{\mu},$$

$$f(R, \tau) = F'/F, \; \nu = \int_0^\infty f(R, \tau) \, d\tau.$$

The function $f(R, \tau)$ is the energy-time integrand, normalized by its peak value at the shock front, expressed as a function of R and a reduced time τ which normalizes its initial slope to -1 if μ does not vanish.

At the shock front, $\boldsymbol{l}_\varphi \times (\boldsymbol{l}_z \cdot \nabla_0 \boldsymbol{r}) = \boldsymbol{n} \sin \vartheta$, where ϑ is the angle between the tangent to the wavefront and the r-axis. The desired energy-equation is obtained by eliminating μ between the first two of equations (h) and making use of the Hugoniot relations, equations (2). One obtains equation (5) as the result.

Central Experiment Station, U. S. Bureau of Mines,
Pittsburgh, Pa.
California Institute of Technology,
Pasadena, Calif.

[6] Our dissipation assumption breaks down if the first shock wave can be overtaken by second shocks built up in its rear. This will not be the case if the pressure-time curve is initially monotone decreasing with asymptotic value p_0. If the excess pressure p' has a negative phase, the second shock will develop in the negative part of the pressure-time curve but cannot overtake the initial positive shock. In this case, our theory will apply to the positive phase if the time integrals are extended to the time at which the excess pressure in the positive phase vanishes.

THE METHOD OF CHARACTERISTICS IN THE THREE-DIMENSIONAL STATIONARY SUPERSONIC FLOW OF A COMPRESSIBLE GAS

BY

N. COBURN AND C. L. DOLPH[1]

1. **Introduction.** Though considerable is known about supersonic flow problems of compressible gases which depend upon two independent variables (for example, stationary plane flows, non-stationary flows in a tube, and so on), little is known about flow problems for more independent variables. Therefore, for mathematical simplicity, we have limited the following discussion to the general stationary, supersonic, isentropic, compressible flow of a perfect gas in three dimensions. However, our methods and ideas admit of application to other flow problems (for example, non-stationary two-dimensional flows, and so on).

Although the theory of characteristics had its origin in the work of Monge in the early nineteenth century, its full significance was first realized by H. Lewy[2] as late as 1928 for the case of two independent variables. Lewy showed that the introduction of the two families of characteristic curves along a given initial non-characteristic curve amounts to the introduction of a new coordinate system in which the differential equation assumes a simple form. This simplicity results from the fact that the original partial differential equation can be replaced by two partial differential equations in each of which there occurs only derivatives with respect to one of the characteristic parameters. Then, Lewy replaced the original differential equation by a system of differential equations consisting of the above two equations, the two equations obtained by factoring the characteristic condition, and the two strip conditions. The method of Lewy was generalized by E. W. Titt[3] to the case of three independent variables, in a paper whose significance for the study of the supersonic flow of compressible gases seems to have been overlooked. The main difficulty in the case of three independent variables is that the characteristic condition is a quadratic form in three variables and of rank three. Hence, the characteristic quadratic form can no longer be factored. However, by use of an analytical scheme, Titt succeeded in showing that two families of characteristic surfaces can be associated with a given initial surface (the characteristic surfaces are not uniquely determined). Now, through each point, there pass two bicharacteristic curves of such nature that each curve lies on one of the characteristic surfaces through the point. Evidently, these curves determine two congruences in the physical

[1] The authors are indebted to the U. S. Air Force for the financial aid which has made this study possible.

[2] Courant-Hilbert, *Methoden der Mathematischen Physik*, vol. 2, 1937, Chap. V.

[3] E. W. Titt, Ann. of Math. (2) vol. 40 (1939) p. 862.

space. Next, Titt showed that these two congruences determine a family of ∞^1 surfaces. Finally, by introducing a new coordinate system consisting of the two families of characteristic surfaces and the family of surfaces through their corresponding bicharacteristics and by requiring that this last family become the surfaces z = constant, Titt found that the three-dimensional problem reduced to a two-dimensional problem. At this point, Lewy's ideas can be applied to prove the existence of a solution.

In view of the introduction of the parameterization, z = constant, for the surfaces determined by the bicharacteristics, we shall be unable to use Titt's resulting equations. The fundamental ideas of our method will be: (1) to consider the ∞^3 surfaces consisting of the two families of characteristic surfaces and the third family of surfaces through the bicharacteristics as forming a natural coordinate system for the problem; (2) to determine the components of the velocity vector with respect to this coordinate system (this process is equivalent to Lewy's factorization of the characteristic condition and Titt's analytical scheme); (3) to determine the components of the directional derivatives of the velocity vector in this coordinate system. By use of this last device, we obtain two equations which determine the form of the original differential equation in terms of the characteristic parameters. Lewy and Titt carry out the corresponding process in their theories by use of determinants; the resulting equations are often called the second characteristic conditions. Thus, we obtain a system of non-linear first order equations, which should reduce to Titt's equations (for the case studied here) if one parameterizes the third family of surfaces as z = constant. It follows that a solution of this system exists and satisfies the original equation. Presumably, finite difference method would be used to compute flows. Finally, we have determined various interesting properties of special flows.

C. Ferrari[4] attempted to carry out a similar procedure for the supersonic flow past a body of revolution with an angle of attack different from zero. Essentially, without proof, he set up a curvilinear coordinate system consisting of two characteristic surfaces through each of the circles around the body of revolution and the set of meridian planes through the axis of symmetry of the body. Unfortunately, except for axial symmetric flows, the meridian planes will not contain the appropriate bicharacteristic directions, that is, will not be the surfaces z = constant. So that one can only conclude that the method of Ferrari would lead to a usable approximation only for very small angles of attack.

2. **Geometry of the flow equation.** In our future work, we shall use the notation of tensor calculus. Subscripts will denote covariant quantities and superscripts will denote contravariant quantities. Further, all indices will run over the range 1, 2, 3. Finally, we shall use the summation convention.

Let us denote by x^λ ($\lambda = 1, 2, 3$) the general coordinates of a Euclidean space with metric tensor $g_{\lambda\mu}$. Further, let v^λ denote the components of the velocity

[4] C. Ferrari, Atti della Reale Accademie della Scienze de Torino vol. 72 (1936) p. 140.

vector of the flow, and let c denote the local sound speed. We define a symmetric tensor $a^{\lambda\mu}$ by the equation

$$a^{\lambda\mu} = v^{\lambda}v^{\mu} - c^{2}g^{\lambda\mu}. \tag{2.1}$$

If ∇_{μ} denotes covariant differentiation, then the basic partial differential equation of the flow is[5]

$$a^{\lambda\mu}\nabla_{\lambda}v_{\mu} = 0. \tag{2.2}$$

For irrotational flows, a velocity potential Φ exists such that

$$v_{\lambda} = \nabla_{\lambda}\Phi. \tag{2.3}$$

It can be easily shown that for irrotational supersonic flows, (2.2) is a hyperbolic equation. The characteristic normal cone is defined by

$$a^{\lambda\mu}i_{\lambda}i_{\mu} = 0, \tag{2.4}$$

where i_{λ} is a unit normal vector which is orthogonal to a family of characteristic surfaces, α = constant.

We shall prove as a first result that the *projection of the velocity vector on the normal to the characteristic surfaces has the magnitude* c *(the local sound speed).* This result is a generalization of the corresponding property for the case of plane flows and implies that the *velocity vector is the axis of the normal cone.*

The result follows immediately upon substituting (2.1) into (2.4). We find

$$(v^{\lambda}i_{\lambda})^{2} = c^{2}. \tag{2.5}$$

Henceforth, we shall consider $v^{\lambda}i_{\lambda} = c$. This choice of sign is equivalent to working with one nappe of the cone (2.4).

It is well known that corresponding to any normal to a characteristic surface, there exists a bicharacteristic direction lying on the characteristic surface. Next, we shall show that *this bicharacteristic direction lies in the plane determined by the normal to the characteristic surface and the velocity vector.*

Let q denote the magnitude of the velocity vector v^{λ}. Consider an arbitrary point of a characteristic surface. Let t^{λ} denote a unit vector lying long the intersection of a plane through i^{λ} and v^{λ} with the characteristic surface. Since the projection of v^{λ} on i^{λ} is c, it follows that the projection of v^{λ} on t^{λ} is $(q^{2} - c^{2})^{1/2}$. Hence, we may write

$$v^{\lambda} - ci^{\lambda} = (q^{2} - c^{2})^{1/2}t^{\lambda}. \tag{2.6}$$

[5] This equation reduces to $(u^2 - c^2)u_x + (v^2 - c^2)v_y + (w^2 - c^2)w_z + uv(u_y + v_x) + uw(u_z + w_x) + vw(v_z + w_y) = 0$ in Euclidean rectangular coordinates. Hence (3.2) is the proper tensor form of the basic potential equation expressed in terms of velocity components. This can also be justified independently by eliminating from the tensor forms of the equations of motion and continuity.

Now, the bicharacteristic directions are defined by the equation[6]

$$\frac{dx^\lambda}{dt} = a^{\lambda\mu} \nabla_\mu \alpha = \underset{\alpha}{M} a^{\lambda\mu} i_\mu, \tag{2.7}$$

where t is the parameter along the bicharacteristic and M_α is the magnitude of the vector $\nabla_\mu \alpha$. By use of (2.1), (2.5), we find that (2.7) reduces to

$$\frac{dx^\lambda}{dt} = \underset{\alpha}{M} c(v^\lambda - ci^\lambda). \tag{2.8}$$

With the aid of (2.6), we can write this last equation as

$$\frac{dx^\lambda}{dt} = \underset{\alpha}{M} c(q^2 - c^2)^{1/2} t^\lambda . \tag{2.9}$$

This verifies our result. We note that since t^λ is in the bicharacteristic direction, (2.6) implies that the *velocity vector is the axis of the cone of bicharacteristics.*

We shall introduce, now, a natural coordinate system by a procedure equivalent to that used by Titt. Consider an arbitrary surface (the initial surface) along which the velocity vector is defined. Introduce any family of ∞^1 curves with unit tangent vector l^λ on this given surface subject to the condition that a plane perpendicular to l^λ shall cut the normal cone in two distinct directions. Thus, at each point of the initial surface, two distinct normals to characteristic surfaces are defined. By considering the strip condition[7] as a partial differential equation for a characteristic surface and knowing two strips along each curve l^λ in the initial surface (the normals to these strips are the directions of intersection of the normal cone and the plane perpendicular to l^λ), two families of characteristic surfaces are defined throughout the flow space. Evidently, the two families of characteristic surfaces intersect the initial surface in the curves l^λ. We now define l^λ as the unit vector of a congruence of curves (in space) which are the intersections of the two families of characteristic surfaces. Further, we note that the unit normals i^λ (for α = constant) and $'i^\lambda$ (for β = constant) to the characteristic surfaces and the corresponding bicharacteristics t^λ and $'t^\lambda$ define four additional congruences in space.

Our next step will be to determine the components of the vectors i^λ, $'i^\lambda$, v^λ with respect to the triad of unit vectors t^λ, $'t^\lambda$, l^λ. In order to do this, we shall determine what relations exist between the angles of the triad. Let

$$a = \cos\psi = l^\lambda t_\lambda, \qquad b = \cos\psi' = l^\lambda{}' t_\lambda, \qquad d = \cos\phi = t^\lambda{}' t_\lambda . \tag{2.10}$$

Since i^λ is orthogonal to t^λ, l^λ; and $'i^\lambda$ is orthogonal to $'t^\lambda$, l^λ, we may write

$$i^\lambda = \frac{\epsilon^{\lambda\alpha\beta} l_\alpha t_\beta}{g^{1/2}(1-a^2)^{1/2}}, \qquad 'i^\lambda = \frac{-\epsilon^{\lambda\alpha\beta} l_\alpha {}'t_\beta}{g^{1/2}(1-b^2)^{1/2}}, \tag{2.11}$$

where $\epsilon^{\lambda\alpha\beta}$ is the permutation symbol and g is the determinant of the metric

[6] See reference of footnote 2, p. 355.

[7] See reference of footnote 2, p. 67 or pp. 82–83.

tensor. The minus sign is used in the second equation of (2.11) since i^λ, $'i^\lambda$ both belong to the same nappe of the normal cone. By use of (2.6) and (2.11), we obtain the relation

$$\frac{c}{g^{1/2}(1-a^2)^{1/2}}\epsilon^{\lambda\alpha\beta}l_\alpha t_\beta + (q^2-c^2)^{1/2}t^\lambda = -\frac{c}{g^{1/2}(1-b^2)^{1/2}}\epsilon^{\lambda\alpha\beta}l_\alpha{}'t_\beta + (q^2-c^{2\prime})^{1/2}{}'t^\lambda. \tag{2.12}$$

Forming the scalar product of (2.12) with l_λ, $'t_\lambda$, respectively, we find

$$t^\lambda l_\lambda = {}'t^\lambda l_\lambda, \qquad \frac{\epsilon^{\lambda\alpha\beta}l_\alpha t_\beta{}'t_\lambda}{g^{1/2}} = \frac{(q^2-c^2)^{1/2}(1-d)(1-a^2)^{1/2}}{c}. \tag{2.13}$$

The first relation of (2.13) implies that $\psi = \psi'$. The left-hand member of the second relation in (2.13) is the volume of the parallelopiped whose sides are l^λ, t^λ, $'t^\lambda$. By an elementary calculation, it can be shown that this second relation is equivalent to the following *relation between* ψ *and* ϕ

$$1 - a^2 = \sin^2\psi = \frac{(1-\cos\phi)c^2}{2c^2-(q^2-c^2)(1-\cos\phi)}. \tag{2.14}$$

By use of the relations (2.13), one can easily determine expressions for the vectors i^λ, $'i^\lambda$, v^λ in terms of t^λ, $'t^\lambda$, l^λ and the scalars q, c, d ($d = \cos\phi$). Thus, for i^λ, we may write

$$i^\lambda = \alpha\,{}'t^\lambda + \beta l^\lambda + \gamma\, t^\lambda, \tag{2.15}$$

where α, β, γ are undetermined scalars. By forming the scalar products of (2.15) with $'t^\lambda$, l^λ, t^λ, we find that the determinant of the coefficients of the right-hand side is $(\epsilon^{\lambda\alpha\beta}l_\alpha t_\beta{}'t_\lambda/g^{1/2})^2$. Using (2.13), we obtain

$$i^\lambda = \frac{c}{(1-d)(q^2-c^2)^{1/2}}\,{}'t^\lambda - \frac{ac}{(1-a^2)(q^2-c^2)^{1/2}}\,l^\lambda + \frac{(a^2-d)c}{(1-a^2)(1-d)(q^2-c^2)^{1/2}}\,t^\lambda, \tag{2.16}$$

where a and d are related by (2.14). A similar computation furnishes

$$'i^\lambda = \frac{(a^2-d)c}{(1-a^2)(1-d)(q^2-c^2)^{1/2}}\,{}'t^\lambda - \frac{ac}{(1-a^2)(q^2-c^2)^{1/2}}\,l^\lambda + \frac{c}{(1-d)(q^2-c^2)^{1/2}}\,t^\lambda, \tag{2.17}$$

$$v^\lambda = \frac{c^2}{(1-d)(q^2-c^2)^{1/2}}(t^\lambda + {}'t^\lambda) - \frac{ac^2}{(1-a^2)(q^2-c^2)^{1/2}}\,l^\lambda. \tag{2.18}$$

Equation (2.18) corresponds to Titt's equations ((2.7), (2.8), (2.9)).

Finally, we shall determine the form of (2.2) when the directional derivatives along t^λ, $'t^\lambda$, l^λ are used. The resulting equations are often called the second characteristic conditions. First, we decompose $\nabla_\mu v_\lambda$ with respect to the directions i^λ, t^λ, l^λ and $'i^\lambda$, $'t^\lambda$, l^λ. That is, we write

$$\nabla_\mu v_\lambda = i_\mu a_\lambda + t_\mu b_\lambda + l_\mu c_\lambda = 'i_\mu \bar{a}_\lambda + 't_\mu \bar{b}_\lambda + l_\mu \bar{c}_\lambda \tag{2.19}$$

where a_λ, b_λ, c_λ, $\bar{a}_\lambda$, $\bar{b}_\lambda$, $\bar{c}_\lambda$ are unknown vectors. A computation shows that

$$\begin{aligned} a_\lambda &= i^\alpha \nabla_\alpha v_\lambda, \qquad b_\lambda = \frac{-t^\alpha \nabla_\alpha v_\lambda + a l^\alpha \nabla_\alpha v_\lambda}{a^2 - 1}, \\ c_\lambda &= \frac{a t^\alpha \nabla_\alpha v_\lambda - l^\alpha \nabla_\alpha v_\lambda}{a^2 - 1}, \end{aligned} \tag{2.20}$$

and similar expressions are obtained for the barred vectors. By multiplying (2.19) by $a^{\lambda\mu}$ and noting that $\nabla_\mu v_\lambda$ is symmetric (that is, $t^\lambda i^\alpha \nabla_\alpha v_\lambda = i^\lambda t^\alpha \nabla_\alpha v_\lambda$, and so on), we obtain with the aid of (2.16), (2.17)

$$\left[\frac{c^2}{1-d}('t^\lambda - t^\lambda) + (q^2 - c^2)^{1/2} v^\lambda\right] t^\alpha \nabla_\alpha v_\lambda + \frac{c^2}{a^2 - 1}[l^\lambda - a t^\lambda] l^\alpha \nabla_\alpha v_\lambda = 0, \tag{2.21}$$

$$\left[\frac{c^2}{1-d}(t^\lambda - 't^\lambda) + (q^2 - c^2)^{1/2} v^\lambda\right] 't^\alpha \nabla_\alpha v_\lambda + \frac{c^2}{a^2 - 1}[l^\lambda - a t^\lambda] l^\alpha \nabla_\alpha v_\lambda = 0. \tag{2.22}$$

3. The system of partial differential equations. Let us introduce a coordinate system (with coordinates α, β, γ) such that the characteristic surfaces are α = constant, β = constant and the surfaces determined by the two congruences of bicharacteristics are γ = constant. The congruences l^λ, t^λ, $'t^\lambda$ become the coordinate lines. Let the first fundamental form for such a set of coordinates be

$$ds^2 = A^2\, d\alpha^2 + B^2\, d\beta^2 + C^2\, d\gamma^2 + 2\,E\, d\alpha\, d\beta + 2\,F\, d\alpha\, d\gamma + 2\,G\, d\beta\, d\gamma, \tag{3.1}$$

and let x, y, z be rectangular Euclidean coordinates. The expressions for A^2, B^2, C^2, E, F, G in terms of the partial derivatives of x, y, z with respect to α, β, γ are well known. Further, the components of the vectors t^λ, $'t^\lambda$, l^λ are $B^{-1}(\partial x/\partial\beta, \partial y/\partial\beta, \partial z/\partial\beta)$, $A^{-1}(\partial x/\partial\alpha, \partial y/\partial\alpha, \partial z/\partial\alpha)$, $C^{-1}(\partial x/\partial\gamma, \partial y/\partial\gamma, \partial z/\partial\gamma)$, respectively. A simple computation and use of (2.10) shows that $a = G/BC, d = E/AB$. The equation (2.14) may be expressed in terms of $A, B, C,$ etc. Finally, we note that

$$t^\alpha \nabla_\alpha v_\lambda = B^{-1}\frac{\partial v_\lambda}{\partial\beta}, \qquad 't^\alpha \nabla_\alpha v_\lambda = A^{-1}\frac{\partial v_\lambda}{\partial\alpha}, \qquad l^\alpha \nabla_\alpha v_\lambda = C^{-1}\frac{\partial v_\lambda}{\partial\gamma}. \tag{3.2}$$

If in the future we write $x = x^1$, $y = x^2$, $z = x^3$, then equation (2.18) becomes

$$\begin{aligned} v^\lambda = {} & \frac{c^2}{(AB - E)(q^2 - c^2)^{1/2}}\left[A\frac{\partial x^\lambda}{\partial\beta} + B\frac{\partial x^\lambda}{\partial\alpha}\right] \\ & - \frac{c^2 BG}{(B^2C^2 - G^2)(q^2 - c^2)^{1/2}}\frac{\partial x_\lambda}{\partial\gamma}. \end{aligned} \tag{3.3}$$

Further, the equations (2.21), (2.22) become

$$(3.4)\qquad \frac{1}{B}\left[\frac{c^2AB}{AB-E}\left(\frac{1}{A}\frac{\partial x^\lambda}{\partial \alpha}-\frac{1}{B}\frac{\partial x^\lambda}{\partial \beta}\right)+(q^2-c^2)^{1/2}v^\lambda\right]\frac{\partial v_\lambda}{\partial \beta} + \frac{c^2B^2C}{G^2-B^2C^2}\left[\frac{1}{C}\frac{\partial x^\lambda}{\partial \gamma}-\frac{G}{B^2C}\frac{\partial x^\lambda}{\partial \beta}\right]\frac{\partial v_\lambda}{\partial \gamma}=0,$$

$$(3.5)\qquad \frac{1}{A}\left[\frac{c^2AB}{AB-E}\left(\frac{1}{B}\frac{\partial x^\lambda}{\partial \beta}-\frac{1}{A}\frac{\partial x^\lambda}{\partial \alpha}\right)+(q^2-c^2)^{1/2}v^\lambda\right]\frac{\partial v_\lambda}{\partial \alpha} + \frac{c^2B^2C}{G^2-B^2C^2}\left[\frac{1}{C}\frac{\partial x^\lambda}{\partial \gamma}-\frac{G}{ABC}\frac{\partial x^\lambda}{\partial \alpha}\right]\frac{\partial v_\lambda}{\partial \gamma}=0.$$

Finally, the symmetry conditions (2.3) furnish the relation

$$(3.6)\qquad \frac{\partial x^\lambda}{\partial \alpha}\frac{\partial v_\lambda}{\partial \beta}=\frac{\partial x^\lambda}{\partial \beta}\frac{\partial v_\lambda}{\partial \alpha}$$

and two others obtained by replacing (α, β) by (α, γ) and by (β, γ). *The equations* (3.3), (3.4), (3.5), (3.6) *constitute a system of* 8 *nonlinear first order partial differential equations for the* 6 *independent variables* x^λ, v^λ $(\lambda = 1, 2, 3)$ *as functions of* α, β, γ. *This system is equivalent to that used by Titt and hence*: (1) *the system is consistent*; *and* (2) *the system furnishes sufficient conditions to determine a solution of* (2.2). Of course, in the above relations $q^2 = v^\lambda v_\lambda$ and q, c are related by the Bernoulli relation $q^2 + (2/(\gamma - 1))c^2 = \bar{q}^2$.

A considerable simplification occurs in the above system when $\psi = \pi/2$ $(a = 0)$. Evidently, $F = G = 0$. Further (2.14) reduce to $d = \cos\phi = (q^2 - 2c^2)/q^2$. The equations (3.3), (3.4), (3.5) become

$$(3.7)\qquad v^\lambda = \frac{q^2}{2(q^2-c^2)^{1/2}}\left(\frac{1}{A}\frac{\partial x^\lambda}{\partial \alpha}+\frac{1}{B}\frac{\partial x^\lambda}{\partial \beta}\right),$$

$$(3.8)\qquad \frac{1}{B}\left[\frac{q^2}{2}\left(\frac{1}{A}\frac{\partial x^\lambda}{\partial \alpha}-\frac{1}{B}\frac{\partial x^\lambda}{\partial \beta}\right)+(q^2-c^2)^{1/2}v^\lambda\right]\frac{\partial v_\lambda}{\partial \beta}-\frac{c^2}{C^2}\frac{\partial x^\lambda}{\partial \gamma}\frac{\partial v_\lambda}{\partial \gamma}=0,$$

$$(3.9)\qquad \frac{1}{A}\left[\frac{q^2}{2}\left(\frac{1}{B}\frac{\partial x^\lambda}{\partial \beta}-\frac{1}{A}\frac{\partial x^\lambda}{\partial \alpha}\right)+(q^2-c^2)^{1/2}v^\lambda\right]\frac{\partial v_\lambda}{\partial \alpha}-\frac{c^2}{C^2}\frac{\partial x^\lambda}{\partial \gamma}\frac{\partial v_\lambda}{\partial \gamma}=0.$$

It should be noted that by choosing l^λ orthogonal to v^λ on the initial surface, the condition $\psi = \pi/2$ $(a = 0)$ is satisfied over the initial surface. Physically, it seems that the condition $\psi = \pi/2$ should then be satisfied throughout the flow space. To date, we have been unable to obtain a proof of this conjecture.

4. **Generalizations of two-dimensional flows—case for which $\psi = \pi/2$.** In this case, at each point, l^λ is perpendicular to the plane containing the vectors t^λ, $'t^\lambda$, i^λ, $'i^\lambda$, v^λ. Since l^λ is perpendicular to v^λ, we obtain our first result: *a necessary and sufficient condition that* $\psi = \pi/2$ *is that the congruence of curves which are the intersections of the two families of characteristic surfaces shall lie on the equipotential surfaces.*

By an analysis of the strip conditions, we shall determine some conditions for degenerate flows. First, we note that for $\psi = \pi/2$, the equations (2.18), (2.21), (2.22) become

$$(4.1) \qquad v^\lambda = \frac{q^2}{2(q^2 - c^2)^{1/2}} (t^\lambda + {}'t^\lambda),$$

$$(4.2) \qquad i^\lambda = \frac{2c^2 - q^2}{2c(q^2 - c^2)^{1/2}} t^\lambda + \frac{q^2}{2c(q^2 - c^2)^{1/2}} {}'t^\lambda,$$

$$(4.3) \qquad {}'i^\lambda = \frac{2c^2 - q^2}{2c(q^2 - c^2)^{1/2}} {}'t^\lambda + \frac{q^2}{2c(q^2 - c^2)^{1/2}} t^\lambda.$$

From the strip conditions,[7] we have

$$(4.4) \qquad \frac{d}{dt} \nabla_\lambda \alpha = -\frac{1}{2} (\nabla_\lambda a^{\mu\nu})(\nabla_\mu \alpha)(\nabla_\nu \alpha),$$

$$(4.5) \qquad \frac{d}{dt'} \nabla_\lambda \beta = -\frac{1}{2} (\nabla_\lambda a^{\mu\nu})(\nabla_\mu \beta)(\nabla_\nu \beta),$$

where α = constant, β = constant are the families of characteristic surfaces and t and t' are the parameters induced along the corresponding bicharacteristics. By use of (2.9), we see that the arc length parameters s, s' along the congruences of bicharacteristics t^λ, ${}'t^\lambda$ are related to the induced parameters t, t' by

$$(4.6) \qquad ds = \underset{\alpha}{M} c(q^2 - c^2)^{1/2}\, dt, \qquad ds' = \underset{\beta}{M} c(q^2 - c^2)^{1/2}\, dt',$$

where M_α, M_β are the magnitudes of the vectors $\nabla_\lambda \alpha$, $\nabla_\lambda \beta$ respectively. Since $i_\lambda = \nabla_\lambda \alpha / M_\alpha$, ${}'i_\lambda = \nabla_\lambda \beta / M_\beta$, we find by use of (4.4), (4.5), (4.6)

$$(4.7) \qquad \underset{\alpha}{M} c(q^2 - c^2)^{1/2} t^\mu \nabla_\mu (\underset{\alpha}{M} i_\lambda) = -\frac{1}{2} \underset{\alpha}{M}{}^2 (\nabla_\lambda a^{\mu\nu}) i_\mu i_\nu,$$

$$(4.8) \qquad \underset{\beta}{M} c(q^2 - c^2)^{1/2} {}'t^\mu \nabla_\mu (\underset{\beta}{M} {}'i_\lambda) = -\frac{1}{2} \underset{\beta}{M}{}^2 (\nabla_\lambda a^{\mu\nu}) {}'i_\mu {}'i_\nu.$$

By expanding (4.7) (or 4.8) and resolving the resulting equation into components along i^λ, t^λ, l^λ, we obtain

$$(4.9) \qquad t^\mu \nabla_\mu \underset{\alpha}{M} = -\frac{\underset{\alpha}{M}}{2c(q^2 - c^2)^{1/2}} (\nabla_\lambda a^{\mu\nu}) i_\mu i_\nu i^\lambda,$$

$$(4.10) \qquad t^\mu t^\lambda \nabla_\mu i_\lambda = -\frac{1}{2c(q^2 - c^2)^{1/2}} (\nabla_\lambda a^{\mu\nu}) i_\mu i_\nu t^\lambda,$$

$$(4.11) \qquad l^\lambda t^\mu \nabla_\mu i_\lambda = -\frac{1}{2c(q^2 - c^2)^{1/2}} (\nabla_\lambda a^{\mu\nu}) i_\mu i_\nu l^\lambda.$$

From (2.1), it follows that

$$(4.12) \qquad \nabla_\lambda a^{\mu\nu} = v^\mu \nabla_\lambda v^\nu + v^\nu \nabla_\lambda v^\mu - g^{\nu\mu} \nabla_\lambda c^2.$$

Further, it is well known that $\nabla_\mu i_\lambda$ can be decomposed into

$$\nabla_\mu i_\lambda = i_\mu u_\lambda + h_{\mu\lambda}, \tag{4.13}$$

where u_λ is the curvature vector of the i_λ congruence and $h_{\mu\lambda}$ is the second fundamental tensor of the characteristic surfaces, α = constant. By use of the last two relations, (4.9), (4.10), (4.11) reduce to

$$t^\mu \nabla_\mu \underset{\alpha}{M} = -\frac{\underset{\alpha}{M}}{(q^2 - c^2)^{1/2}} (i^\lambda i^\mu \nabla_\mu v_\lambda - i^\lambda \nabla_\lambda c), \tag{4.14}$$

$$h_{\mu\lambda} t^\lambda t^\mu = -\frac{1}{(q^2 - c^2)^{1/2}} (t^\lambda i^\mu \nabla_\lambda v_\mu - t^\lambda \nabla_\lambda c), \tag{4.15}$$

$$h_{\mu\lambda} t^\lambda l^\mu = \frac{1}{(q^2 - c^2)^{1/2}} (l^\lambda i^\mu \nabla_\lambda v_\mu - l^\lambda \nabla_\lambda c). \tag{4.16}$$

By use of (4.2) and the special coordinate system used in §3, the right-hand sides of (4.14) through (4.16) can be expressed in terms of partial derivatives of v_λ, c with respect to α, β, γ. Particular interest is attached to the cases in which the left-hand side of (4.14) through (4.16) vanish. Thus, if $t^\mu \nabla_\mu M_\alpha$ vanishes, we have a generalization of one aspect of simple waves (M_α is constant along the two-dimensional simple wave). If $h_{\mu\lambda} t^\mu t^\lambda$ vanishes, then t^λ is an asymptotic direction on the surfaces, α = constant; if $h_{\mu\lambda} t^\lambda l^\mu$ vanishes then t^λ, l^λ are the principal directions on the surfaces, α = constant. Further, if both $h_{\mu\lambda} t^\lambda t^\mu$, $h_{\mu\lambda} t^\lambda l^\mu$ vanish then t^λ are the generators of the developable surfaces, α = constant. In this last case, it follows that i_λ form a parallel vector field along the generators t^λ and hence $t^\mu \nabla_\mu M_\alpha$ must vanish. By expressing the right-hand sides of (4.14) through (4.16) in terms of partial derivatives, we obtain the necessary and sufficient conditions for the validity of the above discussed cases. Each such case furnishes a degenerate type of flow.

Although space limitations do not permit us to furnish details, it can be shown by use of the pair of equations (4.7), (4.8) that: *a necessary condition that a flow be such that l^λ is orthogonal to t^λ, $'t^\lambda$ is that the component of the curvatures of the curves t^λ, $'t^\lambda$ in the l^λ direction be equal.* This condition is obviously satisfied in the cases of plane and axial symmetric flows. In the case of plane flows, t^λ, $'t^\lambda$ are plane curves lying in parallel planes and the l^λ congruence consists of a family of straight lines orthogonal to these planes; in the case of axial symmetric flows, t^λ, $'t^\lambda$ are plane curves lying in meridian planes and the l^λ congruence consists of a family of circles about the axis of the flow.

In order to understand the relation between our equations (2.18) (or (4.1)), (2.21), (2.22) and the usual plane flow equations, it will be desirable to show that *for the case of plane flows, the relation* (2.6) (*which is the basis of* (4.1)) *is equivalent to the equations obtained by factoring* (2.4); *and the relations* (2.21), (2.22) *reduce to the well known second characteristic conditions.*

Evidently, we may write for the case of Euclidean orthogonal coordinates x, y, z, the following formulas: $t^1 = B^{-1}\partial x/\partial\beta$, $t^2 = B^{-1}\partial y/\partial\beta$, $t^3 = 0$;

$i^1 = -B^{-1}\partial y/\partial\beta$, $i^2 = B^{-1}\partial x/\partial\beta$, $i^3 = 0$; $t^\alpha\nabla_\alpha v_1 = B^{-1}\partial u/\partial\beta$, $t^\alpha\nabla_\alpha v_2 = B^{-1}\partial v/\partial\beta$, $v_3 = 0$. By factoring, the equation (2.4) furnishes the following possible solutions for the ratio $\sigma = (\partial y/\partial\beta)/(\partial x/\partial\beta)$

$$\sigma_\pm = \frac{-uv \pm c(q^2 - c^2)^{1/2}}{c^2 - u^2} = \frac{c^2 - v^2}{-uv \mp c(q^2 - c^2)^{1/2}}. \tag{4.17}$$

We shall now show that (4.17) can be obtained by vector operations on (2.6). By forming the scaler product of (2.6) with t^λ, i^λ and dividing the resulting equations, we obtain

$$\sigma' = \frac{v(q^2 - c^2)^{1/2} - cu}{cv + u(q^2 - c^2)^{1/2}}, \qquad \sigma' = \frac{\partial y}{\partial\beta}\Big/\frac{\partial x}{\partial\beta}. \tag{4.18}$$

A simple computation shows that $\sigma' = \sigma_+$. It is clear that the second root σ_- may be obtained by use of the vectors $'t^\lambda$, $'i^\lambda$.

In order to see that (2.21) reduces to the second characteristic relation, we require that (a) shall vanish and that $d = (q^2 - 2c^2)/q^2$. Further, if we eliminate $'t^\lambda$ by use of (2.16), we find that (2.21) reduces to

$$[2c(q^2 - c^2)^{1/2} i^\lambda + (q^2 - 2c^2)t^\lambda]t^\mu\nabla_\mu v_\lambda - c^2 l^\lambda l^\mu \nabla_\mu v_\lambda = 0. \tag{4.19}$$

Since l^λ is orthogonal to v^λ, we find

$$l^\lambda l^\mu \nabla_\mu v_\lambda = -v^\lambda l^\mu \nabla_\mu l_\lambda = -v^\lambda \bar{u}_\lambda, \tag{4.20}$$

where $\bar{u}_\lambda$ is the curvature vector of the l^λ congruence. For plane flows, the l^λ curves are straight lines and hence the last term of (4.19) vanishes. Replacing t^λ, i^λ by their expressions in terms of partial derivatives, we find that (4.19) may be written as

$$-\frac{\partial v}{\partial\beta}\Big/\frac{\partial u}{\partial\beta} = \frac{-2c(q^2 - c^2)^{1/2}\sigma_+ + (q^2 - 2c^2)}{2c(q^2 - c^2)^{1/2} + (q^2 - 2c^2)\sigma_+}. \tag{4.21}$$

Let us write (4.21) in the form

$$-\frac{\partial v}{\partial\beta}\Big/\frac{\partial u}{\partial\beta} = \frac{c^2 - u^2}{c^2 - v^2}\sigma_+ + b, \tag{4.22}$$

where b is unknown, for the present. The usual form of the second characteristic relation is (4.22) with $b = 0$. We shall now show that because of (4.21), (4.17), the term b must vanish in (4.22). Equating (4.21), (4.22), we obtain

$$\begin{aligned}(c^2 - u^2)\sigma_+^2 - 2c(q^2 - c^2)^{1/2}\sigma_+ - (c^2 - v^2)\\ + b\left(\frac{c^2 - v^2}{q^2 - 2c^2}\right)[2c(q^2 - c^2)^{1/2} + \sigma_+(q^2 - 2c^2)] = 0.\end{aligned} \tag{4.23}$$

Eliminating the coefficient $(c^2 - u^2)$ in the first term of (4.23) and the coefficient $(c^2 - v^2)$ in the third term of (4.23) by use of (4.17), we find that b must vanish. Thus, (4.21) is equivalent to (4.22) with $b = 0$. This completes our proof.

In a similar manner, it can be shown that *in the case of axial symmetric flows,* (2.21) *reduces to the second characteristic condition.* We shall omit the proof.

Finally, we shall discuss some geometric properties of the l^λ congruence. Since l^λ lies along the intersection of the characteristic surfaces, α = constant, β = constant and the equipotential surfaces, Φ = constant, we may write

$$\tilde{M}l^\lambda = \frac{\epsilon^{\lambda\mu\nu}}{g^{1/2}} (\nabla_\mu \alpha)(\nabla_\nu \beta), \tag{4.24}$$

$$\bar{M}l^\lambda = \frac{\epsilon^{\lambda\mu\nu}}{g^{1/2}} (\nabla_\mu \alpha)(\nabla_\nu \Phi), \tag{4.25}$$

$$\bar{\bar{M}}l^\lambda = \frac{\epsilon^{\lambda\mu\nu}}{g^{1/2}} (\nabla_\mu \beta)(\nabla_\nu \Phi), \tag{4.26}$$

where $\tilde{M}$, $\bar{M}$, $\bar{\bar{M}}$ are Jacobi Multipliers.[8] Their values can be computed in terms of M_α, M_β. For from the meaning of the vector products in the right-hand sides of the above, we obtain

$$\tilde{M} = M_\alpha M_\beta \frac{2c(q^2 - c^2)^{1/2}}{q}, \tag{4.27}$$

$$\bar{M} = M_\alpha (q^2 - c^2)^{1/2}, \tag{4.28}$$

$$\bar{\bar{M}} = M_\beta (q^2 - c^2)^{1/2}. \tag{4.29}$$

From Jacobi's theorem and Poincare's theorem[9] on integral invariants, it follows that *for the case* $\psi = \pi/2$: (1) l^λ *satisfies the equations* $\nabla_\mu(\tilde{M}l^\mu) = \nabla_\mu(\bar{M}l^\mu) = \nabla_\mu(\bar{\bar{M}}l^\mu) = 0$; (2) *the surfaces,* $\tilde{M}/\bar{M}$ = *constant,* $\bar{M}/\bar{\bar{M}}$ = *constant uniquely determine the* l^λ *congruence*; (3) *the integrals* $\iiint \tilde{M}\, dx\, dy\, dz$, $\iiint \bar{M}\, dx\, dy\, dz$, $\iiint \bar{\bar{M}}\, dx\, dy\, dz$ *are integral invariants of the fluid motion in the* l^λ *direction.* Some interesting conclusions follow if M_α is a function of c only. For such generalized simple waves, it follows that q must be constant along a given l^λ curve.

In concluding, we show that M_α, M_β are related to the metric coefficients A, B. Our result is: $AM_\alpha = BM_\beta = q^2/2c(q^2 - c^2)^{1/2}$.

From the properties of the gradient, we see that for a displacement in the t^λ direction

$$\frac{d\beta}{ds} = t^\lambda \nabla_\lambda \beta. \tag{4.30}$$

[8] The following use of the Theorem of Jacobi was suggested to us by some unpublished results of I. Opatowski on the existence of stream functions for special three dimensional flows. These results were presented to the American Mathematical Society at the University of Chicago on April 25, 1947.

[9] See, for example, Goursat-Hedrick, *Mathematical Analysis*, vol. 2, 1917, pp. 81–86.

But by definition of M_β as the magnitude of $\nabla_\lambda \beta$, we have

$$(4.31) \qquad t^\lambda \nabla_\lambda \beta = \underset{\beta}{M}\, t^{\lambda\prime} i_\lambda = \underset{\beta}{M}\, \frac{2c(q^2 - c^2)^{1/2}}{q^2}.$$

Further, for a displacement in the t^λ direction ($d\alpha = d\gamma = 0$), we find from (3.1) that $ds = B\, d\beta$. Substituting (4.31) into (4.30) and using the relation between ds and $d\beta$ furnishes the desired relation between B and M_β. A similar result is valid for A and M_α.

UNIVERSITY OF MICHIGAN,
ANN ARBOR, MICH.

THE NUMERICAL SOLUTION OF THE TURBULENCE PROBLEM

BY

HOWARD W. EMMONS

Most problems of fluid mechanics have yielded to the application of the general conservation principles (mass, momentum, and energy) so that more and more accurate prediction has been possible.

The stability of laminar flows as computed appears to be properly described by these principles as expressed by the Navier-Stokes Equation [2; 4]. The further growth of turbulence and the description (statistical or otherwise) of the completely developed turbulent motion of fluids has not yet been reduced to the same basic principles. Is this because the principles are inadequate or because the power of our mathematics is inadequate?

The mechanism of transformation from laminar to turbulent motion was first studied experimentally by O. Reynolds [3]. All attempts to calculate the turbulent motion however have resulted only in sets of equations with too many unknowns which, therefore, can only be solved by additional assumptions *not shown to follow from the conservation principles.*

A method of avoiding this difficulty by direct numerical solution of the partial differential equations of the problem was described at the Harvard Symposium on Large Scale Computing Machinery [1].

Briefly, this method proceeds as follows. The Navier-Stokes and continuity equations are used in the form:

$$\zeta_t + \psi_y\zeta_x - \psi_x\zeta_y = \nu\nabla^2\zeta, \qquad \nabla^2\psi = -\zeta. \tag{1}$$

These are written for a two-dimensional flow. A three-dimensional treatment would be analogous. In dimensionless, finite difference form equations (1) become:

$$\text{(a)}\quad \delta^2\{\zeta_0(t+\delta t) - \zeta_0(t) = (4/\text{Re})(\zeta_1 + \zeta_2 + \zeta_3 + \zeta_4 - 4\zeta_0)$$

$$(2)\qquad\qquad - (\psi_4 - \psi_2)(\zeta_1 - \zeta_3) + (\psi_1 - \psi_3)(\zeta_4 - \zeta_2)\},$$

$$\text{(b)}\quad \psi_1 + \psi_2 + \psi_3 + \psi_4 - 4\psi_0 = -\delta^2\zeta_0,$$

where $\delta t = 4\delta^2$, $\text{Re} = (\bar{u}D)/\nu$, and the subscripts refer to values of the variable at the spacial net of points of Fig. 1. The boundary conditions to be satisfied by ψ and ζ are those necessary to insure zero velocity on the surface of all bodies and the appropriate velocities at $\pm\infty$.

A solution proceeds as follows. Some arbitrary ζ distribution is assumed which is consistent with the boundary conditions. This can be taken from the laminar flow solution or in an arbitrary way. (At the initial instant an arbitrary ζ distribution, except on boundary points, may be assumed. The boundary point values are sufficient to satisfy the boundary conditions as seen by equation (2b).)

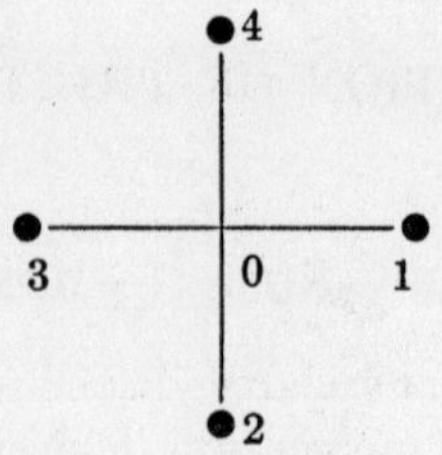

FIG. 1

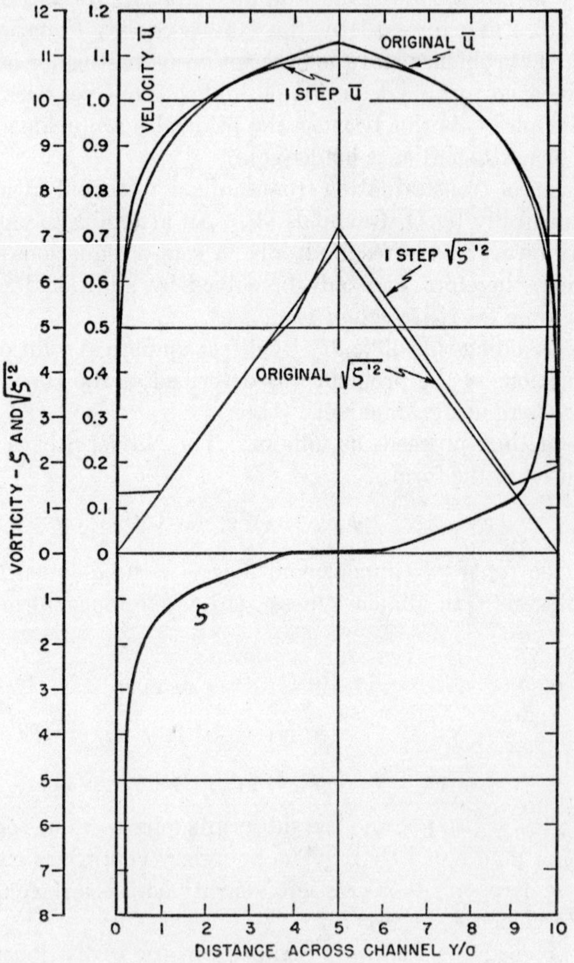

FIG. 2

These initial ζ values together with equations (2b) serve to determine the ψ values at all points at the initial instant. Now equation (2a) is used to find the ζ values at a time δt later. By a repetition of this process the development of $\psi(x, y)$ in time will be determined; that is, a value $\psi(x, y, t)$ at each space-time net point will be found.

If the solution is to be like the physically obtained results, the $\psi(x, y, t)$ would not be expected to have much direct significance. The significant properties of a turbulent flow are various statistical properties. For example, the short time and space average should correspond to the response of a hot-wire ane-

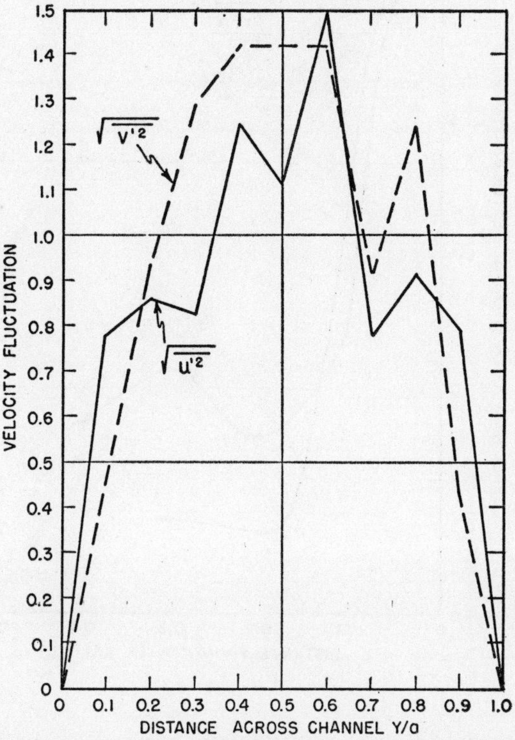

Fig. 3

mometer. A longer time average should correspond to the response of a pitot tube.

A few preliminary calculations have been made.

A piece of two-dimensional channel with 10 net points from wall to wall and 24 net points long has been numerically solved for one time interval. This channel section was repeated to form an "infinite" periodic channel. The solution was started with a mean $\bar{\zeta}$, distribution appropriate to the turbulent flow in a pipe and a fluctuation component, ζ^1, distributed in a gaussian manner with $\overline{\zeta'^2}$ made linear with the distance from the nearest wall. The calculations were made for Re = 4000. After carrying the computation through one time step, the $\bar{\zeta}$ and $\overline{\zeta'^2}$ had changed slightly. See Fig. 2. The mean velocity along the channel $\bar{u}$ changed to a smooth curve as would be expected. The original velocity distribution was chosen as the 1/7 power of the distance from the wall as empirically correct for turbulent flow in a pipe. This law is not correct near the channel center where a discontinuous slope occurs. This is smoothed by the calculations.

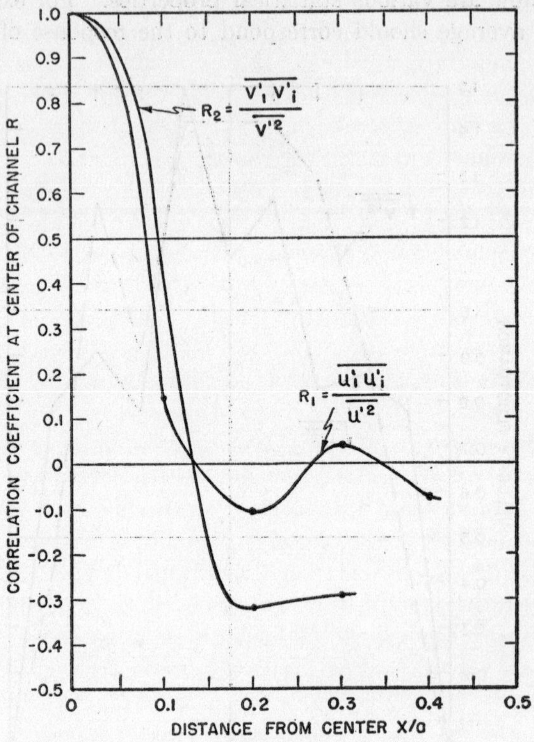

FIG. 4

In Fig. 3, appropriate mean velocity fluctuations are shown:

$$\overline{u'^2}(y) = \frac{1}{24\delta} \sum_{n=0}^{23} (\psi_4(x + n\delta, y, t + \delta t) - \psi_0(x + n\delta, y, t + \delta t))^2,$$

$$\overline{v'^2}(y) = \frac{1}{24\delta} \sum_{n=0}^{23} (\psi_1(x + n\delta, y, t + \delta t) - \psi_0(x + n\delta, y, t + \delta t))^2.$$

Fig. 4 shows the correlation of the velocity at a point i with that at a point j which is distant n from it. That is:

$$R_1 = \frac{\overline{u_1'u_i}}{\overline{u'^2}} = \frac{1}{\overline{u'^2}} \frac{1}{24\delta} \sum_{j=0}^{23} (\psi_4(x + j\delta, y, t + \delta t) - \psi_0(x + j\delta, y, t + \delta t))$$
$$\cdot(\psi_4(x + (n + j)\delta, y, t + \delta t) - \psi_0(x + (n + j)\delta, y, t + \delta t)).$$

This one time increment computation cannot be considered as an adequate treatment in any way but is indicative of several things. The mean properties are similar to those to be expected. The root mean square vorticity fluctuation assumed is not large enough. It has grown almost everywhere. It was also evident during the calculations that the first term on the right of equation (2) was not large enough relative to the second one to dissipate the vorticity produced at the walls.

The further computation of this problem is too tedious to carry out by hand. A large scale sequence-controlled calculating machine is necessary. Such solutions should be carried out for several values of the Reynolds number, Re, so as to cover both laminar and turbulent cases.

Proper perspective must be maintained relative to the computed flow and the experimentally obtained flow. It is known that the instabilities first to develop in a two-dimensional laminar flow are two-dimensional. It is further known, however, that the completely developed turbulent flow is three-dimensional. In the completely developed turbulent flows, therefore, the vortex tubes are stretched as well as convected. This stretching produces a higher dissipation (first term on right of equation (2a)) and therefore greater fluctuations of vorticity would be expected in the two-dimensional case.

There seems, however, to be no reason to suppose that the differential equation system (1) does not possess two types of solution depending upon the value of Re and of the computational disturbance magnitude (that is, the finite difference errors and truncation errors). One type of solution, certainly obtainable at low enough Re, will be exactly the two-dimensional laminar flow solution. The other type of solution obtained at sufficiently high Re should have time-independent statistical mean values although the instantaneous values will be random. The actual values obtained may not agree in magnitude with experimental data, but this will not detract from the value of the solution in showing the possibility of treatment of this type of problem where a unique solution may exist to equations (1) but would have no significance except in its mean properties. For equations (2) as solved, no unique solution would exist because of continual introduction of small random finite difference and truncation errors. In spite of these errors the statistical mean properties must be independent of time if the Navier-Stokes equations and the continuity equations are adequate descriptions of nature.

Bibliography

1. H. W. Emmons, *A new approach to the turbulence problem*, Annals of the Computation Laboratory, vol. 16, Harvard University Press.

2. C. L. M. H. Navier, *Memoirs sur les lois du movements des fluids*, Mémoires de l'Academie des Sciences vol. 6 (1826) pp. 389–440.

3. O. Reynolds, *On the dynamical theory of incorp. viscous fluids and the determination of the criteria*, Philos. Trans. Roy. Soc. London vol. 186 (1895) pp. 123–164.

4. G. Stokes, *On the theory of internal friction of fluids in motion*, Transactions of the Cambridge Philos. Soc. vol. 8 (1847).

Harvard University,
Cambridge, Mass.

ON THE STABILITY OF TRANSONIC FLOWS

BY

Y. H. KUO

It has been theoretically established that the potential flow of a compressible frictionless fluid can actually accelerate or decelerate smoothly through the sonic line in a transonic field as long as the limiting line is not present. The lack of experimental confirmation, however, leads some investigators to question the existence of such type of flows. A. Kantrowitz [3] and J. P. Brown [1] studied the one-dimensional channel flows. The most interesting result is that, when a disturbance characterized by a compression wave travels upstream in a decelerating field, it approaches a stationary state and the wave becomes a shock. The flow is then concluded to be unstable.

In the present paper, a two-dimensional motion is considered. For practical interest, the problem is restricted to a flow around a thin body and hence can be solved by iteration process. By expanding the potential function, taking the thickness ratio as a parameter, the successive approximations can be determined by solving a set of differential equations subject to the prescribed boundary conditions. The equation for the zero-order term is non-linear. Those for higher order terms are all linear but inhomogeneous.

As a disturbance, a special motion has been assumed. It can be shown that, in the case of a plane wave, if the characteristic curve on which both the disturbance potential and its partial derivatives vanish is chosen as a new variable, there exists a simple exact solution. This gives a velocity that at any moment is a linear function of the coordinate x, say. This solution can therefore be used to define a triangular velocity-wave form, which terminates in a discontinuity at a definite maximum velocity to be determined later. The advantage of such a wave is that it is, from the very beginning, discontinuous. Therefore, the singularity encountered by B. Riemann in his study of propagation of plane wave of finite amplitude can never appear. By tracing the history of the wave form during its propagation upstream, it has been found that the slope of the velocity at the tail of the wave decreases with time and approaches zero as time becomes infinite. The velocity of the wave-head also decreases with time but remains always finite. It is clear that once this wave is introduced, it will disappear from the field after a sufficient length of time.

As for the first approximation where the equation involves functions depending upon the shape of the boundary, the body has to be specified. For its simplicity, a slightly curved wavy surface is chosen. In this case when the free-stream Mach number is 0.9 with a thickness about four percent of the wave-length, supersonic regions are locally established at the crests of the surface, as was found by H. Görtler [2]. The mathematical problem is to solve an inhomogeneous partial differential equation subject to the given boundary conditions. By superposition, a solution can thus be constructed. When the velocity of the disturbance-flow is determined, the Rankine-Hugoniot relations can be applied at the wave-head to give a shock-velocity. Since the shock is curved, the strength is restricted to be very weak so that the change of entropy can be neglected. For physical interest, the history of this wave is again traced and the deformation of the triangular wave, due to the presence of the body, is also evaluated.

Bibliography

1. J. P. Brown, *The stability of compressible flows and transition through the speed of sound,* AAF Technical Report 5410.

2. H. Görtler, *Gasströmungen mit Übergang von Unterschall-zu Überschallgeschwindigkeiten,* Zeitschrift für Angewandte Mathematic und Mechanik vol. 20 (1940) pp. 254–262.

3. A. Kantrowitz, *The formation and stability of normal shock waves in channel flows,* NACA Technical Note 1225.

Cornell University,
Ithaca, N. Y.

STABILITY OF THE LAMINAR BOUNDARY LAYER IN A COMPRESSIBLE FLUID[1]

BY

LESTER LEES

Introduction. By the theoretical studies of Heisenberg, Tollmien, Schlichting, and Lin ([1] to [5]) and the careful experimental investigations of Liepmann [6] and H. L. Dryden and his associates [7], it has been definitely established that the flow in the laminar boundary-layer of a viscous homogeneous incompressible fluid is unstable above a certain characteristic critical Reynolds number. When the level of the disturbances in the free stream is low, as in most cases of technical interest, this inherent instability of the laminar motion at sufficiently high Reynolds numbers is responsible for the ultimate transition to turbulent flow in the boundary-layer. The steady laminar boundary-layer flow would always represent a possible solution of the steady equations of motion, but this steady flow is in a state of unstable dynamic equilibrium above the critical Reynolds number. Self-excited disturbances (Tollmien waves) appear in the flow, and these disturbances grow large enough eventually to destroy the laminar motion.

The question naturally arises as to how the phenomena of laminar instability and transition to turbulent flow are modified when the fluid velocities and temperature variations in the boundary layer are large enough so that the compressibility and conductivity of the fluid can no longer be neglected. The success of Lin in clarifying certain mathematical difficulties in the earlier Tollmien-Schlichting analysis opened the way to a theoretical investigation of the stability of the laminar boundary-layer flow of a gas, in which the compressibility and heat conductivity of the gas, as well as its viscosity, are taken into account. The first part of this work was presented in [8]. The objects of this investigation are: (1) to determine how the stability of the laminar boundary-layer is affected by the free-stream Mach number and the thermal conditions at the solid boundary; and (2) to obtain a better understanding of the physical basis for the instability of laminar gas flows.

The stability of the laminar boundary-layer flow of a gas is analyzed by the method of small perturbations, which was already so successfully utilized for the study of the stability of the laminar flow of an incompressible fluid. (See [5].) By this method a nonsteady gas flow is investigated in which all physical quantities differ from their values in a given steady gas flow by small perturbations that are functions of the time and the space coordinates. This nonsteady flow must satisfy the complete gas-dynamic equations of motion and the same boundary conditions as the given steady flow. The question is whether the nonsteady flow damps to the steady flow, oscillates about it, or diverges from it with time—that is, whether the small perturbations are damped, neutral, or self-excited disturb-

[1] A complete account of this investigation will be found in NACA TN No. 1360, 1947. Explicit definitions of symbols may be found in this report.

ances in time, and thus whether the given steady gas flow is stable or unstable. The analysis is particularly concerned with the conditions for the existence of neutral disturbances, which mark the transition from stable to unstable flow and define the minimum critical Reynolds number.

In order to bring out some of the principal features of the stability problem without becoming involved in hopeless mathematical complications, the solid boundary is taken as two-dimensional and of negligible curvature and the boundary-layer flow is regarded as plane and essentially parallel; that is, the velocity component in the direction normal to the surface is negligible and the velocity component parallel to the surface is a function mainly of the distance normal to the surface. The small disturbances, which are also two-dimensional, are analyzed into Fourier components, or normal modes, periodic in the direction of the free stream; and the amplitude of each one of these partial oscillations is a function of the distance normal to the solid surface, that is,

$$u^{*\prime} = \bar{u}_0^* f(y) e^{i\alpha(x-ct)}.$$

In the study of the stability of the laminar boundary layer, it will be seen that only the local properties of the "parallel" flow are significant. To include the variation of the mean velocity in the direction of the free stream, or the velocity component normal to the solid boundary, in the problem would lead only to higher order terms in the differential equations governing the disturbances, since both of these factors are inversely proportional to the local Reynolds number based on the boundary-layer thickness. (See, for example, [2].)

These considerations lead quite naturally to the study of individual partial oscillations of the form $f(y)e^{i\alpha(x-ct)}$, for which the differential equations of disturbance do not contain x and t explicitly.

Because of the involved character of the mathematical analysis only a few remarks will be made here concerning the mathematical development. The present discussion deals rather with certain physical aspects of the stability problem that serve to illuminate the mathematical results and to give an insight into the mechanism of instability in the laminar boundary-layer.

Briefly, the differential equations governing the disturbances in the boundary-layer take the form of a set of six linear differential equations in six independent variables, with c, α, R and M^2 as parameters. For a laminar boundary layer flow $R \gg 1$. Since exact solutions are almost impossible to obtain, the solutions are developed in both convergent and asymptotic series in inverse powers of αR, and only the first terms are utilized in the calculations. (Wasof has shown recently that these solutions exist and that the procedure is justified rigorously.) The solutions must satisfy the boundary conditions that the components of the velocity fluctuations normal and parallel to the main flow vanish at the solid wall and die out at large distances from the wall. The problem of determining the neutral disturbances marking the limits of stability is then an eigenvalue problem; the reader is referred to the original paper for the mathematical discussion.

Energy balance for neutral disturbances. Some of the important conclusions of the stability analysis can be deduced, at least qualitatively, from a study of the energy balance for a neutral disturbance in the laminar boundary layer. There are two regions in the fluid in which a transfer of energy between the mean flow and the disturbance may occur: (1) the "inner critical layer," at which the phase velocity of the disturbance equals the mean flow velocity; and (2) the thin viscous sub-layer adjacent to the solid wall. (These regions are also the interesting ones in the mathematical analysis.)

(1) *Inner critical layer.* At the inner critical layer where $c = w$, the relative velocity between the mean flow and the disturbance is zero and the net transport of momentum across a plane normal to the flow vanishes. Thus, at the critical layer, the vertical transport of momentum, which at other layers of the fluid is balanced by the horizontal transport, can only be accounted for by the diffusive action of viscosity. To be specific, at the critical layer, the equations of motion become, upon elimination of the pressure,

$$v_c^{*\prime}\left[\frac{d}{dy^*}\left(\bar{\rho}^*\frac{d\bar{u}}{dy}\right)\right]_{\bar{u}^*=c^*} = \text{viscous terms.}$$

Now, it was possible to show that $v_c^{*\prime} \neq 0$ for a solution of the differential equations satisfying the boundary conditions. Therefore, if $[(d/dy^*)(\bar{\rho}^* d\bar{u}^*/dy^*)]_{\bar{u}^*=c^*}$ is not zero, the effects of viscosity cannot be ignored at the critical layer. The effect of viscosity is to produce a phase shift in $u^{*\prime}$ across the layer, and the sign and magnitude of this phase shift is determined by the sign and magnitude of the quantity $[(d/dy)(\rho dw/dy)]_{w=c}$, in non-dimensional form. The corresponding apparent shear stress $\tau_c^* = -\bar{\rho}_c^* \overline{u^{*\prime}v^{*\prime}}$ is given by the expression

$$\tau_c^* = \bar{\rho}_0^*(\bar{u}_0^*)^2 \frac{\alpha}{2}\pi \frac{T_c^2}{(w_c')^3}\left[\frac{d}{dy}\left(\frac{w'}{T}\right)\right]_{w=c}.$$

(The "prime" here denotes differentiation with respect to y.) If the quantity $[(d/dy)(w'/T)]_{w=c}$ is negative, the mean flow absorbs energy from the disturbance at the critical layer; if $[(d/dy)(w'/T)]_{w=c}$ is positive, energy passes from the mean flow to the disturbance. If the quantity $[(d/dy)(w'/T)]_{w=c}$ vanishes, there is no exchange of energy. (In the real compressible fluid the actual thickness of the inner critical layer in which the viscous forces are important is of the order of $1/(\alpha R)^{1/3}$.)

If the quantity $|(d/dy)(w'/T)|$ vanishes for some value of $w > 1 - 1/M_0$, a neutral disturbance free from the effects of viscosity always exists in the limit of infinite Reynolds number (inviscid fluid) and the laminar boundary layer flow is unstable at sufficiently large but finite Reynolds numbers.

(2) *Viscous sub-layer adjacent to the solid wall.* As shown by Prandtl, near the solid wall the effect of viscosity on the boundary layer disturbances is analogous to the case of the flat plate oscillating in its own plane. Because of viscous diffusion the phase of the frictional component of the disturbance velocity $u_{fr}^{*\prime}$ is shifted against the "frictionless" or "inviscid" component $u_{inv}^{*\prime}$ in a thin layer

of fluid of thickness of the order of $(\alpha cR/\nu_1)^{-1/2}$. By continuity, the associated normal component $v_{fr}^{*\prime}$ is of the order of

$$|\,\partial u^{*\prime}/\partial x^*\,|\,(\alpha cR/\nu_1)^{-1/2} \approx |\,u_{inv}^{*\prime}\,|_1\,\alpha^{1/2}(cR/\nu_1)^{-1/2}$$

where $(u_{inv}^{*\prime}/\bar{u}_0^*)_1 = T_1/c$. The corresponding apparent shear stress $\tau_1^* = -\rho_1^*\overline{u^{*\prime}v^{*\prime}}$ is given by the expression

$$\tau_1^* \approx \bar{\rho}_0^*(\bar{u}_0^*)^2\,\frac{T_1}{c^2}\left(\frac{\alpha}{cR/\nu_1}\right)^{1/2}.$$

By the action of this shear stress, disturbances of intermediate wave length absorb energy from the mean flow.

Since the shear stress associated with the destabilizing effect of viscosity near the solid surface and the shear stress near the critical layer act roughly throughout the same region of the fluid, the ratio of the rates of energy transferred (approximately $\int_0^h \tau^*(du^*/dy^*)dy^*$) by the two physical processes is

$$\left|\frac{E_c^*}{E_1^*}\right| \sim \left|\frac{\tau_c^*}{\tau_1^*}\right| \approx \frac{\pi}{2}\,\frac{w_1'\,c}{T_1}\,\frac{T_c^2}{(w_c')^3}\left|\left[\frac{d}{dy}\left(\frac{w'}{T}\right)\right]_{w=c}\right| z^{3/2} = \frac{1}{2}\,|\,v(c)\,|\,z^{3/2}$$

where $z^3 \approx (\alpha R/\nu_c)(c^3/(w_1')^2)$.

Several important conclusions can be drawn at once from this result:

(a) If the quantity $(d/dy)(w'/T)$ is *negative* and sufficiently large when $w = c_1$, say, then the rate at which energy is absorbed by the mean flow near the inner critical layer plus the rate at which the energy of the disturbance is dissipated by viscous action more than counterbalances the destabilizing effect of viscosity near the solid surface. Consequently, a neutral disturbance with the phase velocity $c \geqq c_1$ does not exist; in fact, all disturbances for which $c \geqq c_1$ are damped.

Now, when the main flow outside the boundary-layer is *subsonic*, there is always a range of values of phase velocity $0 \leqq c \leqq c_0$ for which the ratio $|\,E_c^*/E_1^*\,|$ is small enough for neutral and self-excited disturbances to exist for Reynolds numbers greater than a certain critical value. In other words, every subsonic laminar boundary-layer flow is ultimately unstable at sufficiently high Reynolds numbers. When the main flow is *supersonic*, however, the phase velocity of the boundary-layer disturbances must be subsonic relative to the free stream velocity, or $\bar{u}_0^* - c^* < \bar{a}_0^*$ and $c > 1 - 1/M_0$, in order that these disturbances shall die out exponentially with distance from the solid surface. Because of this physical requirement, the possibility exists that for certain thermal conditions at the solid surface, the quantity $[(d/dy)(w'/T)]_{w=c}$ is always large enough negatively so that only damped characteristic disturbances exist at all Reynolds numbers. In other words, when the free stream velocity is supersonic, for certain thermal conditions at the solid wall (shortly to be discussed) the laminar boundary-layer is completely stabilized.

(b) From the energy balance for a neutral boundary-layer disturbance, it follows also that the actual stability limit, or minimum critical Reynolds number, $R_{\theta_{cr(\min)}}$, for any laminar boundary layer flow is largely determined by the distribution of the product of the density and vorticity $\rho(dw/dy)$ across the boundary layer.

An approximate estimate of $R_{\theta_{cr(\min)}}$ is obtained that serves as a criterion for the influence of free-stream Mach number and thermal conditions at the solid surface on laminar stability. The maximum value of the phase velocity c_0 for an undamped disturbance corresponds to the value of c for which $(1 - 2\lambda)v = 0.580$, at which point $R_\theta \approx R_{\theta_{cr(\min)}} = (\text{const.}/\alpha c_0^3)\nu_{c_0}(w_c')^2$. Since $\alpha \approx w_1' c_0(1 - M_0^2(1 - c_0)^2)^{1/2}$, the criterion reads as follows, for zero pressure gradient:

$$R_{\theta_{cr(\min)}} \approx \frac{6}{T_1} \frac{[T(c_0)]^{1.76}}{c_0^4(1 - M_0^2(1 - c_0)^2)^{1/2}},$$

where T is the ratio of the temperature at a point within the boundary layer to free-stream temperature, T_1 is the ratio of temperature at the solid surface to the free-stream temperature and c_0 is the value of c for which $(1 - 2\lambda)v = 0.580$. The functions $v(c)$ and $\lambda(c)$ are defined as follows:

$$v(c) = -\frac{\pi(dw/d\eta)_1 c}{T_1}\left[\frac{T^2}{(dw/d\eta)^3}\frac{d}{d\eta}\left(\frac{1}{T}\frac{dw}{d\eta}\right)\right]_{w=c},$$

$$\lambda(c) = \frac{\eta(dw/d\eta)_1}{c} - 1$$

where η (non-dimensional distance from surface) is $y^*(\bar{u}_0^*/\nu_0^* x^*)^{1/2}$.

Effect of free-stream Mach number and thermal conditions at solid surface on stability of laminar boundary layer.

(a) *Stabilization of laminar boundary layer at supersonic Mach numbers.* From simple physical considerations, or directly from the equations of mean motion for laminar boundary layer flow, it is deduced that withdrawal of heat from the fluid through the solid surface at a sufficient rate produces a convex velocity profile, for which the quantity $(d/dy)(w'/T)$ is always negative. The conclusions drawn from a study of the energy balance for a neutral boundary layer disturbance show that when the velocity of the "free stream" is supersonic, the laminar boundary layer is completely stabilized if the rate at which heat is withdrawn through the solid surface reaches or exceeds a critical value that depends only on the Mach number, the Reynolds number, and the properties of the gas. (Calculations of this critical rate of heat transfer have been carried out in the original paper for a series of Mach numbers from 1.5 to 5.0.)

Under steady supersonic free-flight conditions at high altitude, the heat radiated from the surface is balanced by the heat conducted to the surface from the fluid. Calculations show that for $M_0 > 3$ at 50,000 feet altitude, and for $M_0 > 2$

at 100,000 feet altitude, the rate of heat conduction through the solid surface is greater than the critical rate of heat withdrawal for laminar stability. Thus, under these free-flight conditions the laminar boundary layer flow is completely stabilized in the absence of a positive pressure gradient.

It should be noted that under wind-tunnel test conditions, in which the model is stationary, these radiation-conduction effects are absent, not only because of reradiation from the walls of the wind tunnel but also because the surface temperatures are low—generally of the order of room temperature.

(b) *Minimum critical Reynolds number,* $R_{\theta_{cr(\min)}}$, *or stability limit.* On the basis of the stability criterion and a study of the equations of mean motion, the effect of adding heat to the fluid through the solid surface is to produce an inflection-point velocity profile, for which $(d/dy)(w'/T)$ near the surface is positive, and therefore to reduce $R_{\theta_{cr(\min)}}$ and destabilize the flow, as compared with the flow over an insulated surface at the same Mach number. Withdrawing heat through the solid surface has exactly the opposite effect. The value of $R_{\theta_{cr(\min)}}$ for the laminar boundary-layer flow over an insulated surface decreases as the Mach number increases, and the flow is destabilized, as compared with the Blasius flow at low speeds.

Detailed calculations of the curves of wave number (inverse wave length) against Reynolds number for the neutral boundary-layer disturbances for 10 representative cases of insulated and noninsulated surfaces show that the quantitative effect on stability of the thermal conditions at the solid surface is very large. For example, at a Mach number of 0.70, the value of $R_{\theta_{cr(\min)}}$ is 63 when $T_1 = 1.25$ (heat added to fluid), $R_{\theta_{cr(\min)}} = 126$ when $T = 1.10$ (insulated surface), and $R_{\theta_{cr(\min)}} = 5150$ when $T_1 = 0.70$ (heat withdrawn from fluid). Since $R_x^* \approx 2.25R_\theta^2$, the effect on $R_{x^*_{cr(\min)}}$ is even greater.

(c) *Stability of the laminar boundary-layer flow of a gas with a pressure gradient in the direction of the free-stream.* The results of the present study of laminar stability can be extended to include laminar boundary layer flows of a gas in which there is a pressure gradient in the direction of the free stream. Although further study is required, it is presumed that only the local mean velocity-temperature distribution determines the stability of the local boundary-layer flow. If that should be the case, the effect of a pressure gradient on laminar stability could be easily calculated through its effect on the local distribution of the product of mean density and mean vorticity across the boundary layer.

When the free-stream velocity at the "edge" of the boundary layer is supersonic, by analogy with the stabilizing effect of a withdrawal of heat from the fluid, it is expected that the laminar boundary-layer flow is completely stable at all Reynolds numbers when the negative (favorable) pressure gradient reaches or exceeds a certain critical value that depends only on the Mach number and the properties of the gas. The laminar boundary-layer flow over a surface in thermal equilibrium under free flight conditions should be completely stable for $M_0 > M_s$, say, where $M_s < 3$ if there is a negative pressure gradient in the direction of the free stream.

Bibliography

1. Werner Heisenberg, *Über Stabilität und Turbulenz von Flüssigkeitsströmen*, Annalen der Physik (4) vol. 74 (1924) pp. 577–627.

2. W. Tollmien, *Über die Entstehung der Turbulenz*, Nachr. Ges. Wiss. Göttingen (1929) pp. 21–44. (Available as NACA TM No. 609, 1931.)

3. ———, *Ein allgemeines Kriterium der Instabilität laminarer Geschwindigkeitsverteilungen*, Nachr. Ges. Wiss. Göttingen, vol. 1 (1935) pp. 79–114. (Available as NACA TM No. 792, 1936.)

4. H. Schlichting, *Amplitudenverteilung und Energiebilanz der kleinen Störungen bei der Plattenströmung*, Nachr. Ges. Wiss. Göttingen, vol. 1 (1935) pp. 47–78.

5. C. C. Lin, *On the stability of two-dimensional parallel flows*. Part I, Quarterly of Applied Mathematics, vol. 3 (1945) pp. 117–142; Part II, vol 3 (1945) pp. 218–234; Part III, vol. 3 (1946) pp. 277–301.

6. Hans W. Liepmann, *Investigations on laminar boundary-layer stability and transition on curved boundaries*, NACA Wartime Report No. W-107, 1943

7. G. B. Schubauer and H. K. Skramstad, *Laminar-boundary-layer oscillations and transition on a flat plate*, NACA Wartime Report No. W-8, April, 1943.

8. Lester Lees and Chia Chiao Lin, *Investigation of the stability of the laminar boundary layer in a compressible fluid*, NACA TN No. 1115, 1946.

Princeton University,
Princeton, N. J.

REMARKS ON THE SPECTRUM OF TURBULENCE

BY

C. C. LIN

1. Introduction. The theory of isotropic turbulence as developed by von Kármán and Howarth[1] leads to an equation for the propagation of the longitudinal double correlation function $f(r)$ which however includes the triple correlation function $h(r)$. It is therefore hardly possible to get more definite results from it without making further assumptions. In 1941, Kolmogoroff[2] introduced the assumption that

$$\{1 - f(r)\}^{1/2} : \{h(r)\}^{1/3} = \text{constant} \tag{1.1}$$

for moderate values of r and reached the conclusion

$$1 - f(r) \sim r^{2/3} \tag{1.2}$$

for a certain range of values of r. Later, Obukhoff[3] studied the spectrum of turbulence, introduced a certain mechanism for the transfer of energy among various wave numbers, and was able to establish a similar result. Recently Onsager,[4] von Weizsäcker,[5] and Heisenberg[6] studied the same problem from a point of view somewhat similar to that of Obukhoff, and obtained similar conclusions, although the details of the analyses are somewhat different from each other. Heisenberg also studied the spectrum at very high frequencies, while the other authors limited themselves to medium frequencies only.

Since G. I. Taylor has shown that there is a Fourier transform relation between the spectrum and the double correlation function, it is apparent that Kolmogoroff's approach to the problem is closely connected with those of the others. Indeed, the equation of von Kármán and Howarth for the propagation of the double correlation function should correspond to an exact equation for the change of the spectrum. The first part of this note is concerned with the deduction of this relation. This relation connects the assumptions made on the transfer term with the corresponding ones made on the triple correlation function $h(r)$. It therefore enables one to examine more closely the assumptions made by the various authors, and it turns out that they are all different from each other.

As a side result, it is noted that the equation for the change of spectrum shows that the low frequency end of the spectrum is invariant in the course of time. It

[1] Th. von Kármán and L. Howarth, Proc. Roy. Soc. London Ser. A vol. 164 (1938) pp. 192–215.

[2] A. N. Kolmogoroff, C. R. (Doklady) Acad. Sci. URSS vol. 32 (1941) pp. 16–18.

[3] A. Obukhoff, C. R. (Doklady) Acad. Sci. URSS vol. 32 (1941) pp. 19–21.

[4] L. Onsager, Physical Reviews vol. 68 (1945) p. 286 (Abstract only).

[5] C. F. von Weizsäcker, Zeitschrift für Physik vol. 124 (1948) p. 614.

[6] W. Heisenberg, Zeitschrift für Physik vol. 124 (1948) p. 628; Proc. Roy. Soc. London Ser. A vol. 195 (1948) pp. 402–406.

is possible to show that this relation is equivalent to an earlier result of Loitziansky,[7] which states that

$$\overline{u^2} \int_0^\infty f(r) r^4 \, dr = \text{constant}. \tag{1.3}$$

Together with an assumption of similarity in a rather strong form, this relation leads to a definite index for the power law of decay of turbulence, which was first given by von Kármán and Howarth without a fixed index. The present discussion thus gives an independent deduction and a physical interpretation of the result of Loitziansky.

The last part of this note contains a short discussion on the density of zeroes of the velocity fluctuation. By using some known results of Kac[8] and Rice,[9] it is shown that the number of zeros observed per second should be directly proportional to the wind speed and inversely proportional to the three halves power of the wind speed for the same grid. The constant of proportionality should give some indication about the joint frequency distribution of the velocity fluctuation and its space derivatives.

2. The equation for the change of spectrum. As indicated above, the equation for the change of spectrum can be obtained from the propagation equations of double correlations

$$\frac{\partial}{\partial t}(\overline{u^2} f) + 2(\overline{u^2})^{3/2}\left(\frac{\partial h}{\partial r} + \frac{4h}{r}\right) = 2\nu\overline{u^2}\left(\frac{\partial^2 f}{\partial r^2} + \frac{4}{r}\frac{\partial f}{\partial r}\right), \tag{2.1}$$

where $f(r, t)$ is the longitudinal correlation function at time t for two points at a distance r apart, $\overline{u^2}$ is the mean square value of the velocity fluctuations, $h(r, t)$ is a triple correlation function as defined by von Kármán and Howarth, and ν is the kinematic viscosity coefficient. G. I. Taylor has shown that there is the pair of Fourier transform relations

$$f(r) = \int_0^\infty \cos\frac{2\pi nr}{U} F_0(n)\, dn,$$

$$F_0(n) = \frac{4}{U}\int_0^\infty \cos\frac{2\pi nr}{U} f(r)\, dr,$$

where $F_0(n)\, dn$ is the fraction of energy lying between frequencies n and $n + dn$, and U is the wind velocity. If one considers a spatial decomposition, one would naturally write the above relations in the form

$$f(r) = \int_0^\infty \cos kr F_1(k)\, dk, \quad F_1(k) = \frac{2}{\pi}\int_0^\infty \cos kr f(r)\, dr, \tag{2.2}$$

where k is 2π times the wave number.

[7] See A. N. Kolmogoroff, C. R. (Doklady) Acad. Sci. URSS vol. 31 (1941) pp. 538–540.

[8] M. Kac, Bull. Amer. Math. Soc. vol. 49 (1943) pp. 314–320; Amer. J. Math. vol. 65 (1943) pp. 609–615.

[9] S. O. Rice, Bell System Technical Journal vol. 24 (1945) pp. 51–53.

Applying a Fourier transform operation to (2.1), one then obtains

$$\frac{\partial}{\partial t}\{\overline{u^2}F_1(k)\} + 2(\overline{u^2})^{3/2}\left\{k^2H_1(k) - 4\int_\infty^k kH_1(k)\,dk\right\} = 2\nu\overline{u^2}\left\{-k^2F_1(k) + 4\int_\infty^k kF_1(k)\,dk\right\} \tag{2.3}$$

where $H_1(k)$ is given by the pair of relations

$$h(r) = \int_0^\infty kH_1(k)\sin kr\,dr,\quad kH_1(k) = \frac{2}{\pi}\int_0^\infty h(r)\sin kr\,dr. \tag{2.4}$$

The right-hand side represents the dissipation of energy. It is, however, not in the expected form proportional to $k^2F_1(k)$. This discrepancy is removed if we note that $F_1(k)$ is the spectrum corresponding to a one-dimensional Fourier analysis of the turbulent field, while the isotropic turbulent field should be subjected to a three-dimensional analysis which is related to the one-dimensional spectrum $F_1(k)$ by the relation

$$F(k) = k^2F_1''(k) - kF_1'(k). \tag{2.5}$$

This may be obtained from the relation

$$2F_1(k_x) = F_x(k_x) = \frac{1}{4}\int_{k_x}^\infty \frac{dk}{k^3}(k^2 - k_x^2)F(k), \tag{2.6}$$

given by Heisenberg.[10] Performing the operation indicated by (2.5) on (2.3), we have an equation in the expected form

$$\frac{\partial}{\partial t}\{\overline{u^2}F(k)\} + 2(\overline{u^2})^{3/2}k^2H(k) = -2\nu\overline{u^2}k^2F(k), \tag{2.7}$$

where $H(k)$ is defined by

$$H(k) = 2\{k^2H_1''(k) - kH_1'(k)\}. \tag{2.8}$$

Equation (2.7) is the rigorous equation for the change of spectrum; the term $k^2H(k)$ represents the transfer of energy. The function $H(k)$ must satisfy the condition

$$\int_0^\infty k^2H(k)\,dk = 0$$

which also implies

$$\int_0^\infty k^2H_1(k)\,dk = 0.$$

[10] Loc. cit. Equation (34).

3. **Discussion of the various assumptions.** Obukhoff based his discussions on an equation of the form (2.7) without, however, correlating his transport term with the triple correlation $h(r)$. His assumption is equivalent to

$$\int_k^\infty k^2 H(k)\,dk = 2K\left[\int_0^k k^2 F(k)\,dk\right]^{1/2} \cdot \int_k^\infty F(k)\,dk. \tag{3.1}$$

Von Weizsacker and Heisenberg considered a quantity S_k In Heisenberg's work, this quantity S_k is essentially

$$S_\kappa = -\frac{\partial}{\partial t}\int_0^k \overline{u^2} F(k)\,dk$$

$$= 2\rho(\overline{u^2})^{3/2}\int_0^k k'^2 H(k')\,dk' + 2\mu\overline{u^2}\int_0^k k'^2 F(k')\,dk'.$$

He assumed

$$S_\kappa = 2(\mu + \eta_\kappa)\int_0^k F(k')k'^2\,dk'$$

with

$$\eta_\kappa = \kappa\rho(\overline{u^2})^{1/2}\int_k^\infty \left(\frac{F(k')}{k'^3}\right)^{1/2} dk', \qquad \kappa = \text{constant.}$$

This means that he assumed

$$\int_0^k k'^2 H(k')\,dk = \kappa\int_k^\infty \left(\frac{F(k')}{k'^3}\right)^{1/2} dk' \cdot \int_0^k F(k')k'^2\,dk'. \tag{3.2}$$

In addition, he assumed S_k to be independent of k for large k. In Weizsacker's work, the terms in μ were not considered.

Onsager put

$$\frac{d\overline{u^2}}{dt} = \beta(\overline{u^2})^{3/2} k\{kF(k)\}^{3/2} = \text{constant} \tag{3.3}$$

according to a cascade process. This leads directly to

$$F(k) \sim k^{-5/3}, \tag{3.4}$$

which is also obtained by Obukhoff, Weizsacker and Heisenberg. Kolmogoroff obtained the same result from his assumption (1.1) mentioned above. This assumption cannot be shown to be equivalent to assumptions regarding the spectrum by using (2.2), (2.4), (2.5) and (2.8).

Heisenberg is the only one who gave an expression for $F(k)$ when k is very large. He obtained

$$F(k) \sim k^{-7}. \tag{3.5}$$

A closer examination[11] shows that a similar investigation following Obukhoff's line does not lead to any definite result.

[11] This was carried out at the suggestion of Professor Heisenberg.

It thus appears that the theories of the various authors are not exactly equivalent to one another, and that Heisenberg's form gives the best results.

4. **The behavior for small k.** More definite results can be obtained for small values of k. It is obvious from (2.2) and (2.4) that $F_1(k)$ and $H_1(k)$ behave like

$$
\begin{aligned}
F_1(k) &= \alpha - \beta k^2 + \gamma k^4 + \cdots, \\
H_1(k) &= a - bk^2 + ck^4 + \cdots
\end{aligned}
\tag{4.1}
$$

for small k, where

$$
\begin{aligned}
\alpha &= \frac{2}{\pi}\int_0^\infty f(r)\,dr, \quad 2!\beta = \frac{2}{\pi}\int_0^\infty f(r)r^2\,dr, \quad 4!\gamma = \frac{2}{\pi}\int_0^\infty f(r)r^4\,dr, \cdots, \\
a &= \frac{2}{\pi}\int_0^\infty h(r)\,\gamma\,dr, \quad 3!b = \frac{2}{\pi}\int_0^\infty h(r)r^3\,dr, \quad 5!c = \frac{2}{\pi}\int_0^\infty h(r)r^5\,dr, \cdots.
\end{aligned}
\tag{4.2}
$$

The expansion can be carried out as far as these integrals are convergent. By substituting (4.1) into (2.5) and (2.8), the functions $F(k)$ and $H(k)$ are seen to be of the form

$$
\begin{aligned}
F(k) &= 8\gamma k^4 + \cdots, \\
H(k) &= 8ck^4 + \cdots.
\end{aligned}
\tag{4.3}
$$

Hence, the equation (2.7) shows that the spectrum at low frequencies is not changed. Comparing with (4.2), we see that this is another way of arriving at the relation (1.3) given by Loitziansky.

Evidently, the spectrum of turbulence at low frequencies must depend on the size and shape of the experimental apparatus involved. This tends to complicate matters and makes Loitziansky's result valid only under idealized conditions. It might therefore be expected that the index for the power law of decay would become closer to 5/7 when the size of the apparatus becomes very large compared with the scale of turbulence.

5. **The zeros of velocity fluctuations.** The frequency at which the fluctuating velocity becomes zero gives some information about the spectrum, although a rather incomplete one. The following discussions give some tentative conclusions about these zeros. If $p(\xi, \eta)\,d\xi\,d\eta$ is the joint probability for finding u lying between ξ and $\xi + d\xi$ together with $\partial u/\partial x$ lying between η and $\eta + d\eta$, then the density of the number of zeros of $u(x)$ is[8,9]

$$N = \int_{-\infty}^{\infty} |\eta|\, p(0, \eta)\,d\eta.$$

If $p(\xi, \eta)$ is similar for all the cases, that is,

$$p(\xi, \eta)\,d\xi\,d\eta = p_0\left(\frac{\xi}{(\overline{u^2})^{1/2}}, \frac{\eta}{\left(\overline{\left(\frac{\partial u}{\partial x}\right)^2}\right)^{1/2}}\right) d\,\frac{\xi}{(\overline{u^2})^{1/2}}\, d\,\frac{\eta}{\left(\overline{\left(\frac{\partial u}{\partial x}\right)^2}\right)^{1/2}},$$

where p_0 is the same for all cases, then

$$N = \left\{ \left(\overline{\left(\frac{\partial u}{\partial x} \right)^2} \right)^{1/2} \Big/ (\overline{u^2})^{1/2} \right\} \cdot \int_{-\infty}^{\infty} p_0(0, \eta_0)\, d\eta_0 \mid \eta_0 \mid$$

which is proportional to $1/\lambda$.

If we follow G. I. Taylor in assuming that the field of flow is being carried downstream without any appreciable change, we see that the number of zeros per second is

$$NU \sim U/\lambda.$$

This is to be expected, as λ is a scale measuring the rapidity of change.

Behind similar grids, Taylor has shown that

$$\lambda \sim U^{-1/2}.$$

Hence, the number of zeros per second should be proportional to $U^{3/2}$. This can be checked experimentally.

If the joint distribution function is, say, Gaussian, the constant of proportionality can be explicitly determined. Thus, from the measurement of the density of zeros, one has an idea of the distribution function, although not complete information.[12] Thus, if the proportionality law just given is verified experimentally, there is ground to support the assumption of similar distribution.

BROWN UNIVERSITY,
PROVIDENCE, R. I.

[12] The idea of trying to get information from the distribution of zeros was first mentioned to me by Dr. H. L. Dryden.

TWO-DIMENSIONAL COMPRESSIBLE FLOWS

BY

I. OPATOWSKI

1. **Introduction.** The purpose of this paper is to extend to the field of compressible flows some results in the theory of harmonic functions due to T. Levi-Civita [7, 15]. Those results originated with a note of V. Volterra [17], who considered harmonic functions $\varphi(x^1, x^2)$ in the space, dependent only on two general curvilinear orthogonal coordinates, and defined the stream function of an ideal flow of which $\varphi(x^1, x^2)$ was the velocity potential. T. Levi-Civita determined all the systems of coordinates (x^1, x^2, x^3), not necessarily orthogonal, which have the property that

$$\operatorname{divgrad} f(x^1, x^2, x^3) = F(x^1, x^2, x^3)G[f],$$

where f and F are functions of x^1, x^2, x^3 and $G[\cdots]$ is a linear differential operator whose coefficients are independent of x^3. The solutions $\varphi(x^1, x^2)$ of $G[\varphi] = 0$ are *velocity potentials* of steady ideal flows independent of x^3. From a physical viewpoint it is of more interest to find flows whose *velocity vectors* are independent of one of the coordinates. This investigation is done in the present paper for steady compressible flows of perfect gases. Stream functions are introduced for these flows by combining the classical methods with an application of the last multiplier of Jacobi. The use of the latter is known from the theory of Newtonian potential [1, 9, 11, 13, 14] and from J. Larmor's interpretation of the density of a fluid as a last multiplier of Jacobi [6].

2. **Notations.** We use standard notations of the tensor calculus for the covariant and contravariant components of a vector $\boldsymbol{q}$, that is, q_i and q^i respectively. Parentheses are used to indicate powers of these quantities, for example, $(q_i)^2, (q^i)^2, (q)^2$. Projections of vectors on coordinate axes are indicated by subscripts in parentheses, for example, $q_{(i)}$ is the projection of $\boldsymbol{q}$ on the tangent to the coordinate curve along which x^i varies. The symbol of summation with respect to greek letters, if they appear simultaneously as subscripts and superscripts, is omitted. The metric tensor of general curvilinear coordinates $x^\lambda (\lambda = 1, 2, 3)$ is indicated by $a_{\lambda\mu}$, that is,

$$(ds)^2 = a_{\lambda\mu} dx^\lambda dx^\mu,$$

and the determinant $\| a_{\lambda\mu} \|$ by a. Differentiation is denoted by commas, that is, $\partial\varphi/\partial x^\lambda = \varphi_{,\lambda}$.

3. **Two-dimensional compressible flows.** Let $\boldsymbol{q}$ be the velocity of the fluid, p its pressure, ρ its density, ∇_α the symbol of the covariant derivative and $p'(\rho) = dp/d\rho$, then the equations of steady motions of fluids not subject to external or viscous forces are [16, p. 3; 3, p. 240; 8, p. 306]:

$$q^\alpha \nabla_\alpha q_\lambda = -p'(\rho)\rho_{,\lambda}/\rho. \tag{1}$$

The condition of continuity, $\mathrm{div}(\rho q) = 0$, is:

$$(2) \qquad (q^{\lambda} a^{1/2})_{,\lambda}/a^{1/2} = -q^{\lambda}\rho_{,\lambda}/\rho.$$

Multiplying (1) by q^{λ}, summing with respect to λ and eliminating $\rho_{,\lambda}/\rho$ by means of (2), we obtain:

$$(3) \qquad (q^{\alpha} q^{\lambda} \nabla_{\alpha} q_{\lambda})/p'(\rho) = \mathrm{div}\, \boldsymbol{q}.$$

Consider a polytropic transformation of a perfect gas, then $p/\rho^{m} = \text{const.}$ and $p/\rho = RT$, where m, R are constants and T is the absolute temperature. This gives (cf. [16, pp. 5, 6]):

$$(4) \qquad p'(\rho) = 2^{-1}(m - 1)[(q_{\max})^{2} - (q)^{2}],$$

where $q_{\max}$ is the so-called limiting or maximum velocity along each streamline, i.e.

$$(5) \qquad 2^{-1}(q_{\max})^{2} = m(m - 1)^{-1} R T_{q=0}\,.$$

$T_{q=0}$ is the stagnation temperature. The combination of (3) with (4) gives:

$$(6) \qquad q^{\alpha} q^{\lambda} \nabla_{\alpha} q_{\lambda} = 2^{-1}(m - 1)[(q_{\max})^{2} - (q)^{2}]\, \mathrm{div}\, \boldsymbol{q}.$$

Any vector $\boldsymbol{q}(x^{1}, x^{2}, x^{3})$ which is the gradient of a function $\varphi(x^{1}, x^{2}, x^{3})$ and which satisfies (6) is the velocity vector of a compressible potential flow. For such flow $m = c_p/c_v$, where c_p and c_v are the specific heats of the gas at constant pressure and constant volume respectively. *Consider steady flows for which* $\boldsymbol{q}$ *is independent of* x^{3}. Following the results of T. Levi-Civita [7, 15] we prove the existence of the following three families of such two-dimensional compressible potential flows:

(i) If the metric tensor is independent of x^{3} this variable does not appear in (6) for any $\boldsymbol{q}$ independent of x^{3}. There are only three types of systems of coordinates which have such a metric tensor [7], they are cylindrical, axially symmetric and helical coordinates (cf. [7, 13, 14]). Only the third system gives rise to a new family of compressible flows, the *helical flows*. The latter are obtained by taking $x^{1} = r$, $x^{2} \equiv \nu = z - n\omega$, $x^{3} = \omega$, where (z, r, ω) are the ordinary cylindrical coordinates and n is a constant; $\nu = \text{const.}$ are then helical surfaces with the z-axis as their geometric axis. The following is the most general type of velocity potential φ which gives helical flows, that is, flows with $\boldsymbol{q}$ dependent on r and ν only:

$$\varphi = A\omega + \Phi(r, \nu),$$

where A is an arbitrary constant and Φ is independent of ω. A differential equation for Φ may be obtained from (6) by routine procedures of the tensor calculus. For $n = A = 0$ that equation reduces to the differential equation of axially symmetric compressible potential flows.

(ii) *Conical flows.* The existence of this type of flow is well known. Take

$x^1 = \omega$, $x^2 = \theta$, $x^3 = r$, where (r, θ, ω) are the ordinary spherical coordinates. Since [3, p. 110–114]:

$$q_{(1)} = q_1/(r \sin \theta) = q^1 r \sin \theta, \tag{7}$$
$$q_{(2)} = q_2/r = q^2 r, \qquad q_{(3)} = q^3 = q_3,$$

routine calculations show that if $\boldsymbol{q}$, that is, the $q_{(i)}$'s are independent of r, each side of (6) is a product of $1/r$ by an expression independent of r. Therefore r does not appear in equation (6). The most general type of velocity potential compatible with the independence of the $q_{(i)}$'s from r is $\varphi = r\Phi(\theta, \omega)$, where Φ is independent of r. A differential equation for Φ may be obtained from (6).

One may ask whether there exist compressible conical flows with a *velocity potential* φ independent of r. The answer is positive because transforming (6) into an equation in $\varphi(\omega, \theta)$ by putting $q_i = \varphi_{,i}$ the variable r appears in (6) only as a factor $(r)^{-4}$ common to both sides. However q and consequently also $q_{\max}$ are in such flows inversely proportional to r, which is incompatible with constant stagnation temperature (cf. [5]).

(iii) *Spiral flows.* Consider the system of coordinates $x^1 \equiv \sigma = r/\epsilon$, $x^2 = \theta$, $x^3 = \omega$, where (r, θ, ω) are spherical coordinates, $\epsilon = \exp(n\omega)$ and n is a constant. $\sigma =$ constant are spiral surfaces. The metric tensor is here dependent on ω, each of the $a_{\lambda\mu}$'s containing the factor ϵ^2 [7; 14, p. 493–495]. The Christoffel symbols of the 2nd kind are independent of ω. If $q_{(i)}$ $(i = 1, 2, 3)$ are the projections of q on the coordinate curves (see §2), then [15, 12]:

$$q_i = q_{(i)}(a_{ii})^{1/2}. \tag{8}$$

Consider spiral vector fields, that is, vectors $\boldsymbol{q}$ characterized by the fact that the $q_{(i)}$'s expressed in spiral coordinates are independent of $\boldsymbol{\omega}$. Then, simple calculations based on (8) and on the expressions of the $a_{\lambda\mu}$'s [14, p. 493–495] show that:

$$q_1 = \epsilon q_{(1)}; \qquad q_2 = \epsilon\sigma q_{(2)}; \qquad q_3 = \epsilon\sigma\Theta q_{(3)}. \tag{9}$$

$$q^1 = [(1 + n^2 \csc^2 \theta)q_{(1)} - n\Theta q_{(3)} \csc^2 \theta]/\epsilon; \tag{10}$$
$$q^2 = q_{(2)}/(\epsilon\sigma); \qquad q^3 = (\Theta q_{(3)} - nq_{(1)})/(\epsilon\sigma \sin^2 \theta),$$

where $\Theta = (n^2 + \sin^2 \theta)^{1/2}$. A consequence of (9) and (10) is that q and $q_{\max}$ are independent of ω and that each side of (6) expressed in spiral coordinates is a product of $\exp(-n\omega)$ by an expression independent of ω. This proves the existence of compressible flows with a velocity field which is a spiral vector field and with stagnation temperature which is a constant. The velocity potential of these flows is of the type:

$$\varphi = \Phi(\sigma, \theta) \exp(n\omega),$$

where Φ is independent of ω. A differential equation for Φ may be obtained from

(6). Flows with a *velocity potential* dependent only on (σ, θ) exist also, but imply a stagnation temperature proportional to $\exp(2n\omega)$.

4. Rotational flows. The existence of the types of flows discussed in §3 was proved under the condition of irrotationality. Removal of this restriction does not invalidate the conclusions. This may be seen directly from the following equations of the flow (cf. (1), (4), (2)):

$$q^{\alpha}\nabla_{\alpha}q_{\lambda} = -2^{-1}(m-1)[(q_{\max})^2 - (q)^2]\, d\log\rho/dx^{\lambda}, \tag{11}$$

$$(\rho a^{1/2} q^{\lambda})_{,\lambda} = 0. \tag{12}$$

In fact, if the metric tensor is independent of x^3 and the components $q_{(i)}$ are also independent of x^3, the variable x^3 does not appear in the system (11), (12). The existence of rotational conical flows is well known. For spiral flows the same type of reasoning as the one used in §3 (iii) shows that if the $q_{(i)}$'s and ρ are dependent on σ and θ only, ω does not appear in (11). Besides this, since $a^{1/2}$ contains ω only through a factor $\exp(3n\omega)$, the left-hand side of (12) does not contain ω except for the factor $\exp(2n\omega)$ which drops from the equation.

5. An extension of Larmor's theorem and a generalization of the stream function. It is known that the equation

$$q^{\lambda}(x^1, x^2, x^3)W_{,\lambda}(x^1, x^2, x^3) = 0, \tag{13}$$

in which the q^{λ}'s are known functions of the x^i's, has two independent solutions W, which put equal to arbitrary constants are equations of two families of surfaces. Jacobi [4] has given a method for the determination of the family C of curves which are the intersections of those surfaces, if a solution $M(x^1, x^2, x^3)$ of the equation

$$(Mq^{\lambda})_{,\lambda} = 0$$

together with a solution $W \equiv \gamma$ of (13) are known. The equations of C as given by Jacobi are:

$$\gamma(x^1, x^2, x^3) = \text{const.}, \tag{14}$$

$$\psi(x^1, x^2) \equiv \int M(q^2\, dx^1 - q^1\, dx^2)/\gamma_{,3} = \text{const.}, \tag{15}$$

where in the integrand of ψ the variable x^3 is eliminated by means of (14); the constants in (14), (15) are arbitrary. Larmor [6] pointed out that this theorem has a simple meaning in hydrodynamics of steady flows. If $q(q^1, q^2, q^3)$ is the velocity of the fluid and W a solution of (13), $W = \text{const.}$ is a locus of stream-lines. Therefore the curves determined by (14), (15) are the stream-lines of the flow. The function M, a last multiplier of Jacobi, is obtained from the equation of continuity (12):

$$M = \rho a^{1/2}. \tag{16}$$

Here ρ can be eliminated by means of q and $q_{\max}$ (cf. [16, p. 133]). This expression of M was given by Larmor in the case of cartesian orthogonal coordinates ($a = 1$). Whittaker [18] gives credit for it also to Boltzmann, probably because of his application of the Jacobi's multiplier to the kinetic theory of gases [2]. Since

$$2^{-1}q^2 + \int dp/\rho \tag{17}$$

is constant along a stream-line, if (17) can be expressed in terms of (x^1, x^2, x^3) it can be used as the function γ in (15). The following are types of flows for which γ and ψ can be obtained without using (17).

(i) *Flows characterized by the condition that a contravariant velocity component is zero.* Let $q^3 \equiv 0$, then equation (13) is satisfied by $W = x^3$. Therefore the equations of stream-lines are by (14), (15), (16):

$$\gamma \equiv x^3 = \text{const.}, \tag{18}$$

$$\psi(x^1, x^2) \equiv \int \rho a^{1/2}(q^2\, dx^1 - q^1\, dx^2). \tag{19}$$

This type of flow includes the ordinary cylindrical and axially symmetric flows [16, p. 18–21] as well as *spherical flows.* The latter are characterized in spherical coordinates $x^1 = \omega$, $x^2 = \theta$, $x^3 = r$ by the condition $q^3 = q_{(3)} = 0$; the equation of stream-lines on each sphere $r = \text{const.}$ is by (7), (19):

$$\psi(\omega, \theta)/r \equiv \int \rho q_{(2)} \sin\theta\, d\omega - \rho q_{(1)}\, d\theta = \text{const.} \tag{20}$$

(ii) *Flows characterized by the condition that two contravariant velocity components are of the type:*

$$q^i = f(x^1, x^2, x^3)Q^i(x^1, x^2) \qquad (i = 1, 2); \tag{21}$$

where Q^1 and Q^2 are independent of x^3. Then any function $\gamma(x^1, x^2)$ independent of x^3 and satisfying the equation:

$$Q^1\gamma_{,1} + Q^2\gamma_{,2} = 0 \tag{22}$$

is also a function W satisfying (13). Since a function $N(x^1, x^2)$ such that

$$\gamma_{,1} = NQ^2, \qquad \gamma_{,2} = -NQ^1 \tag{23}$$

exists, the equations of stream-lines can be written in the form:

$$\gamma(x^1, x^2) = \text{const.},$$
$$\psi(x^2, x^3) = \int \rho a^{1/2}[(q^3/Q^2)\, dx^2 - f\, dx^3]/N = \text{const.}, \tag{24}$$

the variable x^1 being eliminated in the integrand by means of $\gamma(x^1, x^2) = \text{const.}$ The above expression of ψ is obtained by permuting the subscripts (1, 2, 3) of

(15) into (2, 3, 1). The following are the flows which are of the general type defined in this section:

(a) *Helical flows.* These are characterized by the condition that the $q_{(i)}$'s expressed in helical coordinates are independent of ω (cf. §3 (i)). Since the metric tensor of helical coordinates is independent of ω, the quantities q^i, q and ρ are independent of ω if the flow is helical. Therefore (2) is (cf. [13, 14]):

$$[\partial(r\rho q^1)/\partial r] + [\partial(r\rho q^2)/\partial \nu] = 0. \tag{25}$$

Since for helical flows $Q^i = q^i$, we can take $N = r\rho$ in (23). In fact, the condition of integrability of

$$\partial\gamma/\partial r = r\rho q^2, \qquad \partial\gamma/\partial\nu = -r\rho q^1 \tag{26}$$

is (25). The stream-lines are the intersections of the surfaces $\gamma(r, \nu) = \text{const.}$ defined by (26) with the following surfaces, defined by the second equation (24):

$$\psi(\nu, \omega) \equiv \omega - \int (q^3/q^2)\, d\nu = \text{const.},$$

where r is eliminated in q^3/q^2 by means of $\gamma(r, \nu) = \text{const.}$

(b) *Conical flows* (§3(ii)). The stream-lines of these flows are the intersections of cones $\gamma(\omega, \theta) = \text{const.}$ with surfaces of revolution $\psi(\theta, r) = \text{const.}$, where $\gamma(\omega, \theta)$ is any solution of the equation (cf. [(22), (7)]):

$$[q_{(1)}(\omega, \theta) \cdot \partial\gamma/\partial\omega] + [q_{(2)}(\omega, \theta) \sin\theta \cdot \partial\gamma/\partial\theta] = 0$$

and ψ is obtained according to general methods, for example, from (24).

(c) *Spiral flows* (§3(iii)). The q^i's are here, by (10), of the type (21) with $f = \exp(-n\omega)$. The flows are constrained to the families of surfaces $\gamma(\sigma, \theta) = \text{const.}$, where γ is the solution of the differential equation (cf. (22), (10)):

$$(\epsilon q^1\sigma \cdot \partial\gamma/\partial\sigma) + (q_{(2)} \cdot \partial\gamma/\partial\theta) = 0;$$

ϵq^1 is here independent of ω by (10). Knowing γ, we can obtain complete equations of the stream-lines by applying the general methods.

Differential equations for most of the stream functions ψ and γ considered in this section can be obtained from (6) by routine calculations.

Acknowledgment. The writer is indebted to Prof. C. P. Wells for bibliographical references and to Prof. N. Coburn for his help in connection with the application of the tensor calculus in the present paper. To the criticism of Prof. C. L. Dolph is due the question, what are the systems of coordinates in which divgrad $f(x^1, x^2)$ has the form indicated in § 1, if $f(x^1, x^2)$ is independent of x^3? This question was answered by T. Levi-Civita [7, 111–112] and was found by him to be equivalent to the a priori more restrictive problem of his stated in § 1.

Bibliography

1. E. Betti, *Teorica delle forze Newtoniane*, Pisa, 1879, pp. 150–160.

2. L. Boltzmann, *Der aus den Sätzen über Wärmegleichgewicht folgende Beweis des Prinzips des letzten Multiplikators in seiner einfachsten Form*, Math. Ann. vol. 42 (1893) pp. 374–376; also Wissenschaftliche Abhandlungen vol. 3 (1909) 497–499.

3. L. Brillouin, *Les tenseurs en mécanique et en élasticité*, New York, Dover, 1946.

4. C. G. J. Jacobi, *Vorlesungen über Dynamik*, 2d ed., Berlin, 1884, pp. 74–79.

5. H. Lamb, *Hydrodynamics*, 6th ed., Cambridge University Press, 1932, p. 108.

6. J. Larmor, *A kinematic representation of Jacobi's theory of the last multiplier*, Reports of the British Association for the Advancement of Science, Toronto, 1897, pp. 562–563; also Mathematical and physical papers, vol. 2, Cambridge University Press, 1929, p. 704.

7. T. Levi-Civita, *Tipi di potenziali che si possono far dipendere da due sole coordinate*, Memorie di R. Accademia delle Scienze di Torino (2) vol. 49 (1900) pp. 105–152.

8. ———, *Der absolute Differentialkalkül*, Berlin, Springer, 1928.

9. E. L. Mathieu, *Théorie du potentiel*, I, Paris, Gauthier-Villars, 1885, pp. 165–175.

10. I. Opatowski, *Sulle linee di forza dei potenziali Newtoniani simmetrici*, Atti di R. Accademia delle Scienze di Torino vol. 68 (1932) pp. 135–146.

11. ———, *Sulle funzioni biarmoniche come prodotti analoghi ai prodotti di Lamé e sulle linee di forza dei campi Newtoniani*, Rendiconti di R. Accademia Nazionale dei Lincei (6) vol. 17 (1933) pp. 1049–1054 and vol. 18 (1933) pp. 18–25.

12. ———, *Sui sistemi di congruenze di curve nelle varietá ad n dimensioni*, Atti di R. Istituto Veneto di Scienze, Lettere ed Arti vol. 93 (1934) pp. 1476–1479.

13. ———, *Sulle generalizzazione della funzione associata e sui potenziali elicoidali.* Atti di I Congresso Unione Matematica Italiana, 1938, pp. 387–390.

14. ———, *Force-lines in Newtonian bidimensional fields. Some applications and generalizations*, Revista de Ciencias vol. 41 (1939) pp. 485–502.

15. G. Ricci and T. Levi-Civita, *Méthodes de calcul différentiel absolu et leurs applications*, Math. Ann. vol. 54 (1901) pp. 191–193.

16. R. Sauer, *Theoretische Einführung in die Gasdynamik*, Berlin, Springer, 1943.

17. V. Volterra, *Sopra alcuni problemi della teoria del potenziale*, Annali di Scuola Normale Pisa, 1883, pp. 210–225.

18. E. T. Whittaker, *Analytical dynamics*, 3d ed., Cambridge University Press, 1927, pp. 278–279.

University of Michigan,
Ann Arbor, Mich.

POLYGONAL APPROXIMATION METHOD IN THE HODOGRAPH PLANE

BY

H. PORITSKY

1. **Introduction.** Very little progress has been made in the direct solution of the non-linear equations of steady two-dimensional flow of a compressible fluid in the physical plane ((2.5), (2.7)). More success has attended the solution of the linear equations of flow in the hodograph plane ((2.22), (2.23)), that is, the plane in which the velocity components u, v are the Cartesian coordinates. Yet, a great deal of difficulty still remains in solving compressible flow problems arising in flight and industry.

It has been pointed out by Chaplygin that the differential equation satisfied in the hodograph plane by the stream function in two-dimensional compressible fluid flow can be reduced to the Laplace equation, provided that the equation of state in the (p, V)-plane $(V = 1/\rho)$ (2.1) be replaced by a straight line. Chaplygin chose this straight line as the tangent to the abiabatic curve at p_0, V_0 corresponding to the stagnation (or impact) pressure of the gas. For an object immersed in a field of uniform flow (for instance for an airfoil in a uniform air stream) von Kármán proposed a straight line tangent to the equation of state at the point (p_∞, V_∞) corresponding to the undisturbed uniform flow at infinity. This is known as the "Kármán-Tsien method", and is widely used in aeronautical engineering.

A natural extension of the above methods consists in approximating the equation of state by means of *not one but several* straight line segments in the $(p, 1/\rho)$-plane. This method was discussed by the author in item [1] of the bibliography at the end of the paper.[1] In the following this method is further extended and illustrated.

2. **Fluid flow equations in the physical and the hodograph planes.** As an introduction to the polygonal approximation method which is explained in the next section, we review briefly the fundamental equations.

Elimination of the pressure from the force and continuity equations for the two-dimensional steady flow of a non-viscous compressible fluid satisfying a relation

$$p = p(\rho) \tag{2.1}$$

leads to the differential equation

$$u_x(1 - u^2/a^2) + v_y(1 - v^2/a^2) - (u_y + v_x)uv/a^2 = 0, \tag{2.2}$$

where subscripts denote partial derivatives, and a is the velocity of sound, given by

$$a^2 = dp/d\rho. \tag{2.3}$$

[1] The writer is indebted to G. Horvay for pointing out that in [1] the c_i should have been replaced by their reciprocals. The error has been corrected in the present paper.

If the flow is irrotational, then a velocity potential φ exists such that

$$u = \varphi_y , \qquad v = \varphi_y . \tag{2.4}$$

Introducing φ in (2.1) leads to the equation

$$\varphi_{xx}(1 - \varphi_x^2/a^2) + \varphi_{yy}(1 - \varphi_y^2/a^2) - 2\varphi_{xy}\varphi_x\varphi_y/a^2 = 0 \tag{2.5}$$

while in terms of the flow function ψ, where[2]

$$\rho u = \psi_y, \qquad \rho v = -\psi_x, \tag{2.6}$$

2.2 becomes[3]

$$\begin{aligned}(\psi_y/\rho)_x(1 - \psi_y^2/\rho^2a^2) - (\psi_x/\rho)_y(1 - \psi_x^2/\rho^2a^2) \\ + (\psi_x\psi_y/\rho^2a^2)[(\psi_y/\rho)_y - (\psi_x/\rho)_x] = 0.\end{aligned} \tag{2.7}$$

Both (2.2) and (2.7) are non-linear equations, their non-linearity arising not merely from the squares and products but also from the quantities a and ρ which are related with p and the velocity magnitude

$$w = (u^2 + v^2)^{1/2} \tag{2.8}$$

by means of the Bernoulli equation

$$w^2/2 + \int_{p_0}^{p} dp/\rho = 0, \tag{2.9}$$

where p_0 is the stagnation pressure at which the gas velocity vanishes. When (2.1) reduces to the adiabatic law

$$p = p_0\rho^\gamma/\rho_0^\gamma \tag{2.10}$$

(2.3), (2.9) become

$$a^2 = \gamma p/\rho, \tag{2.11}$$

$$a^2 + (\gamma - 1)w^2/2 = a_0^2 \tag{2.12}$$

respectively, where a_0 is the velocity of sound at the stagnation conditions p_0, ρ_0,

$$a_0^2 = \gamma p_0/\rho_0 . \tag{2.13}$$

Except for desultory attempts at integrating the equations (2.5), (2.7) by successive approximation methods, by numerical and experimental methods, comparatively little progress has been made in the solution of these equations. Even for a flow around a circular cylinder it is not known up to what Mach number a solution of (2.5) exists.

[2] Very often ψ is defined by $\psi_y = u\rho/\rho_0$, $-\psi_x = v\rho/\rho_0$ where ρ_0 is the "stagnation" density.

[3] This equation, in contrast to (2.5), does not assume that the flow is irrotational.

To translate the problem into the hodograph plane, in which u, v are rectangular coordinates, and w, $\theta = \tan^{-1} v/u$ polar coordinates, write (2.4), (2.6) in the form

$$d\varphi = u\,dx + v\,dy, \qquad d\psi = -\rho v\,dx + \rho u\,dy. \tag{2.14}$$

Solving for dx, dy

$$\begin{aligned} dx &= u\,d\varphi/(u^2 + v^2) - v\,d\psi/\rho(u^2 + v^2) \\ &= \cos\theta\,d\varphi/w - \sin\theta\,d\psi/\rho w, \\ dy &= \sin\theta\,d\varphi/w + \cos\theta\,d\psi/\rho w, \end{aligned} \tag{2.15}$$

then substituting for $d\varphi$, $d\psi$

$$d\varphi = \varphi_w\,dw + \varphi_\theta\,d\theta, \qquad d\psi = \psi_w\,dw + \psi_\theta\,d\theta, \tag{2.16}$$

there results

$$\begin{aligned} dx &= x_w\,dw + x_\theta\,d\theta \\ &= (\cos\theta\varphi_w/w - \sin\theta\psi_w/\rho w)\,dw + (\cos\theta\varphi_\theta/w - \sin\theta\psi_\theta/\rho w)\,d\theta, \\ dy &= y_w\,dw + y_\theta\,d\theta \\ &= (\sin\theta\varphi_w/w + \cos\theta\psi_w/\rho w)\,dw + (\sin\theta\varphi_\theta/w + \cos\theta\psi_\theta/\rho w)\,d\theta. \end{aligned} \tag{2.17}$$

Applying the conditions

$$(x_w)_\theta = (x_\theta)_w, \qquad (y_w)_\theta = (y_\theta)_w \tag{2.18}$$

which render dx, dy perfect differentials, one obtains two equations, which upon multiplication by $\sin\theta$, $\cos\theta$, then by $\cos\theta$, $-\sin\theta$ and addition, yield

$$\frac{\varphi_w}{w} = \left(\frac{1}{\rho w}\right)'\psi_\theta, \qquad \frac{\psi_w}{\rho w} = \frac{\varphi_\theta}{w^2}; \tag{2.19}$$

here $'$ indicates differentiation with respect to w. Taking differentials of (2.9) and eliminating dp by means of (2.3) one obtains

$$\frac{d\rho}{dw} = -\frac{\rho w}{a^2} \tag{2.20}$$

and hence

$$\frac{d(1/\rho w)}{dw} = -\frac{1}{\rho w^2} - \frac{d\rho/dw}{\rho^2 w} = \frac{1}{\rho}\left(\frac{1}{a^2} - \frac{1}{w^2}\right). \tag{2.21}$$

Equations (2.19) may now be rewritten

$$\frac{w}{\rho}\varphi_w = \frac{1}{\rho^2}\left(\frac{w^2}{a^2} - 1\right)\psi_\theta = \frac{1}{\rho^2}(M^2 - 1)\psi_\theta, \quad \frac{w}{\rho}\psi_w = \varphi_\theta, \tag{2.22}$$

where M is the Mach number $= w/a$. Elimination of φ from the equations (2.22) leads to

$$\frac{\partial}{\partial w}\left(\frac{w}{\rho}\psi_w\right) = \frac{w}{\rho}\left(\frac{1}{a^2} - \frac{1}{w^2}\right)\psi_{\theta\theta}. \tag{2.23}$$

For the adiabatic law (2.10), elimination of ρ and a reduces equation (2.23) to the form

$$w^2\psi_{ww} + wF(w)\psi_w + G(w)\psi_{\theta\theta} = 0, \tag{2.24}$$

where

$$F(w) = \frac{1 - ((\gamma - 3)/2)(w^2/a_0^2)}{1 - ((\gamma - 1)/2)(w^2/a_0^2)},$$
$$G(w) = \frac{1 - ((\gamma + 1)/2)(w^2/a_0^2)}{1 - ((\gamma - 1)/2)(w^2/a_0^2)}. \tag{2.25}$$

Product solutions of (2.24) of the form

$$\psi = \cos k\theta w^k Y_k \tag{2.26}$$

where k is a constant were obtained by Chaplygin who pointed out that by introducing

$$\tau = (\gamma - 1)w^2/2a_0^2 \tag{2.27}$$

and substituting (2.26) in (2.24), (2.25), one is led to the hypergeometric differential equation for $Y_k(\tau)$ with coefficients which are functions of k and β.

If (2.1) can be represented by a linear relation between p and $1/\rho$:

$$p = A - B/\rho, \tag{2.28}$$

where A and B are (positive) constants, then (2.24) can be reduced to the Laplace equation in θ and a suitable function of w, $W(w)$ as follows. Equations (2.3), (2.9) now yield

$$dp/d\rho = a^2 = B/\rho^2, \tag{2.29}$$

$$w^2 = a^2 - C = B/\rho^2 - C, \tag{2.30}$$

where C is an appropriate constant, and one obtains

$$\frac{1}{\rho^2}\left(\frac{w^2}{a^2} - 1\right) = -\frac{C}{B} \tag{2.31}$$

so that (2.22) reduce to

$$\frac{w}{\rho}\varphi_w = -\frac{C}{B}\psi_\theta, \qquad \frac{w}{\rho}\psi_w = \varphi_\theta. \tag{2.32}$$

If now W is introduced by means of

$$dW = \left(\frac{C}{B}\right)^{1/2}\frac{\rho}{w}\,dw \tag{2.33}$$

then (2.32) become the Cauchy-Riemann equations in the functions ψ, $(B/C)^{1/2}\varphi$ in the variables W, θ, so that $\psi + i(B/C)^{1/2}\varphi$ becomes an analytic function of $W + i\theta$ (or of any analytic function of it) while at the same time (2.23) reduces to the Laplace equation in W, θ.

Substitution from 2.30 in 2.33 leads to

$$(2.34)\qquad \begin{aligned} W &= \left(\frac{C}{B}\right)^{1/2} \int \frac{\rho}{w}\, dw = C^{1/2} \int \frac{dw}{w(w^2 + C)^{1/2}} \\ &= -\ln \frac{C^{1/2} + (w^2 + C)^{1/2}}{w} + K \end{aligned}$$

where K is a constant.

It is also convenient to introduce the complex variable

$$(2.35)\qquad \zeta = re^{i\theta} = e^{W+i\theta}$$

where

$$(2.36)\qquad r = e^{W}.$$

The explicit relation between r and w is given by

$$(2.37)\qquad r = K' \frac{w}{C^{1/2} + (w^2 + C)^{1/2}} = \frac{K'((w^2 + C)^{1/2} - C^{1/2})}{w},$$

where K' is a constant. The lines $\theta =$ const. in the hodograph plane are transformed into the same lines in the ζ-plane. The radial distance $|\zeta| = r$ is a function of W, or w only. Multiplying $\psi + i(B/C)^{1/2}\varphi$ by i, we put

$$(2.38)\qquad -(B/C)^{1/2}\varphi + i\psi = f(\zeta),$$

thus identifying ψ with the imaginary part of $f(\zeta)$ in (2.38).

It is of interest to note from (2.38) that φ, ψ are *not* conjugate harmonics.

After φ, ψ have been determined in the hodograph plane, one must solve for the flow in the physical plane. Equations (2.15) yield

$$(2.39)\qquad dz = dx + i\,dy = \frac{e^{i\theta}}{w}\left(d\varphi + i\frac{d\psi}{\rho}\right).$$

In case (2.28) holds, utilizing (2.30), (2.37), (2.38), one reduces the above to

$$(2.40)\qquad \begin{aligned} dz &= \frac{e^{i\theta}}{w}\left[-\left(\frac{C}{B}\right)^{1/2} \operatorname{Re}(df) + i\left(\frac{w^2 + C}{B}\right)^{1/2} \operatorname{Im}(df)\right] \\ &= \frac{1}{2B^{1/2}}\,[\zeta\, df/K' - K'\, d\bar{f}/\bar{\zeta}], \end{aligned}$$

where bars denote conjugates.

3. Polygonal approximation to the adiabatic in the $(p, 1/\rho)$-plane. If the adiabatic relation (2.10) is approximated by means of *several* straight lines in the $(p, 1/\rho)$-plane:

$$(3.1)\qquad p = A_i - B_i/\rho, \qquad \rho_{i-1} < \rho < \rho_i,$$

then relations (2.29) with proper values of B_i, C_i hold in each interval. At the transition points ρ_i, w takes on values w_i, yielding circles in the hodograph plane. There the sound velocity a is *discontinuous*, and one obtains from (2.9), noting that $w = 0$ for $\rho = \rho_0$,

$$C_1 = B_1, \qquad C_{i+1} - C_i = (B_{i+1} - B_i)/\rho_i^2 = -\Delta(a^2). \tag{3.2}$$

It is convenient to choose the constants K_i in (2.34) so that the values of W join on continuously, and similarly for the r-values in (2.37), so that one may consider a single (W, θ)-plane and a single ζ-plane in which the vertices p_i, ρ_i correspond to the straight lines W_i, and to circles r_i respectively.

The equations (2.19), (2.22) were obtained by making dx, dy perfect differentials, so that the integrals $\int dx$, $\int dy$ derived from (2.17) are independent of the path of integration and yield x, y as true point functions of w, θ. At $w = w_i$ the independence of these integrals of the path of integration must be examined separately. Here we must make x_θ, y_θ continuous. Now from (2.17) one obtains

$$\begin{aligned} x_\theta &= \frac{\cos\theta}{w}\varphi_\theta - \frac{\sin\theta}{\rho w}\psi_\theta, \\ y_\theta &= \frac{\sin\theta}{w}\varphi_\theta + \frac{\cos\theta}{\rho w}\psi_\theta. \end{aligned} \tag{3.3}$$

The determinant of φ_θ, ψ_θ in (3.2) is equal to $1/\rho w^2$, and it, along with the coefficients of φ_θ, ψ_θ, is continuous at w_i. Hence φ_θ, ψ_θ themselves are continuous across $w = w_i$, and so are φ, ψ. It follows now from (2.32) for each w-interval that ψ_w, $B_i/C_i\, \varphi_w$ are continuous at $w = w_i$. From the continuity of $\psi_w = \partial\psi/\partial w$ and from (2.33) follows that $\psi_W = \partial\psi/\partial W$ is discontinuous at $W = W_i$ and that

$$\left(\frac{C_i}{B_i}\right)^{1/2} \frac{\partial\psi}{\partial W}\bigg|_{W_i-} = \left(\frac{C_{i+1}}{B_{i+1}}\right)^{1/2} \frac{\partial\psi}{\partial W}\bigg|_{W_i+} \tag{3.4}$$

the subscripts W_i^-, W_i^+ in (3.4) referring to left-hand and right-hand derivatives at $W = W_i$ respectively. We shall put (3.4) in the form

$$\frac{\partial\psi}{\partial W}\bigg|_{W_i+} = c_i \frac{\partial\psi}{\partial W}\bigg|_{W_i-}, \qquad c_i = \left(\frac{B_{i+1}C_i}{C_{i+1}B_i}\right)^{1/2}. \tag{3.5}$$

In terms of r this condition becomes

$$\frac{\partial\psi}{\partial r}\bigg|_{r_i+} = c_i \frac{\partial\psi}{\partial r}\bigg|_{r_i-}. \tag{3.6}$$

Similarly, one obtains from the continuity of ψ, ψ_θ,

$$\frac{\partial\varphi}{\partial W}\bigg|_{W_i+} = \frac{1}{c_i}\frac{\partial\varphi}{\partial W}\bigg|_{W_i-}, \tag{3.7}$$

$$\left.\frac{\partial \varphi}{\partial r}\right|_{r_i+} = \frac{1}{c_i}\left.\frac{\partial \varphi}{\partial r}\right|_{r_i-}. \tag{3.8}$$

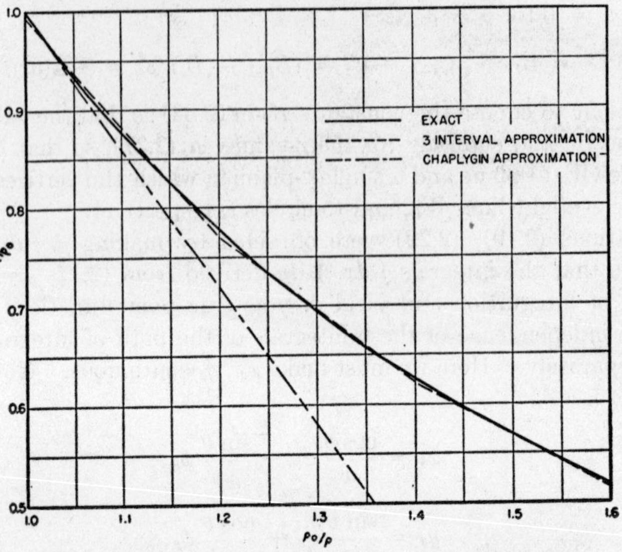

Fig. 1. The p vs. $1/\rho$ curve, and its approximations

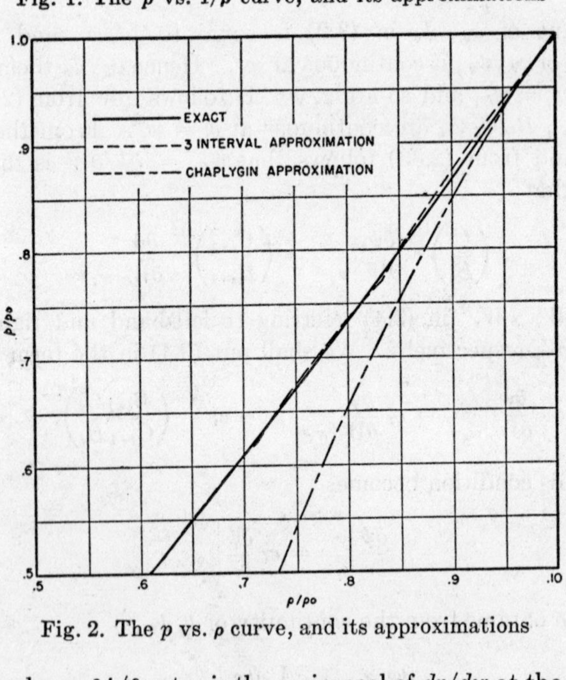

Fig. 2. The p vs. ρ curve, and its approximations

The ratio of the slopes $\partial\psi/\partial r$ at r_i is the reciprocal of dr/dw at the same r_i-values. Thus, on the hodograph plane $\partial\psi/\partial w$ is continuous, while on the ζ-plane, $\partial\psi/\partial r$ is discontinuous at r_i.

The polygonal approximations (3.1) to the adiabatic (2.10) may be viewed in two different ways. One way is to suppose that just enough heat is supplied to or taken away from the gas as it expands so that it obeys equation (3.1); or else one may imagine the existence of a fictitious gas for which the adiabatic reduces to (3.1). In either case the velocity of sound is discontinuous at ρ_i.

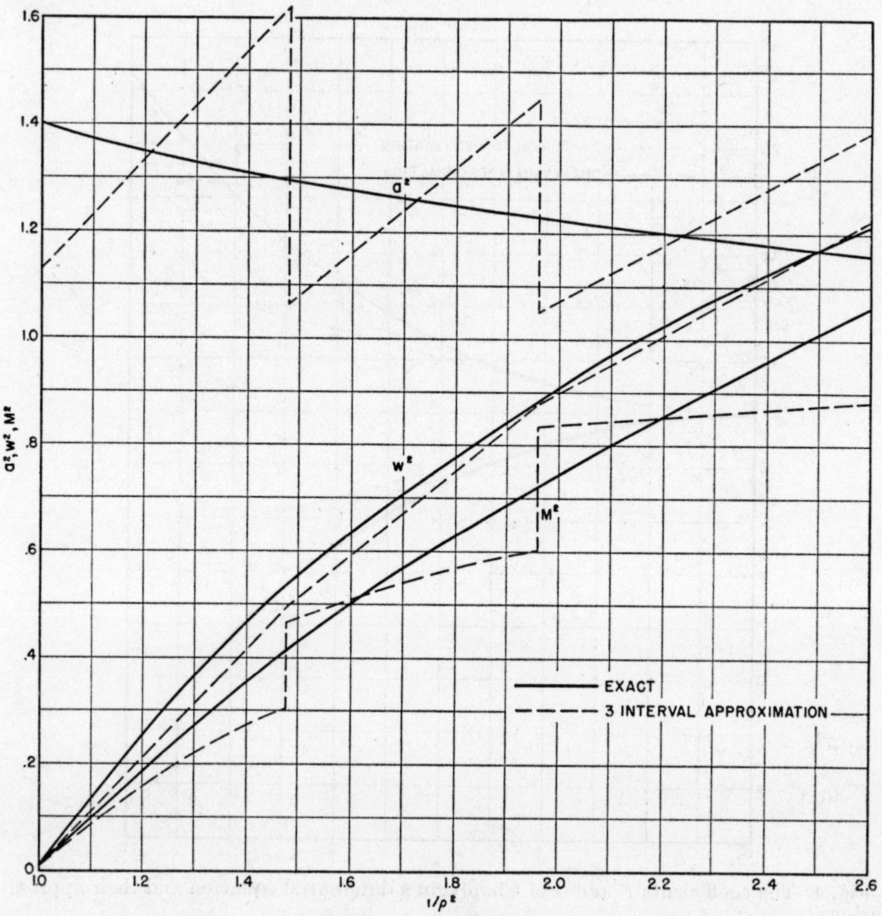

Fig. 3. The curves a^2, w^2, M^2, and their three-interval approximations

As an example, Fig. 1 shows in the $(p, 1/\rho)$-plane the adiabatic relation (2.10) over the range $1 < \rho_0/\rho < 1.6$, a three-interval approximation to it, with $\rho_1/\rho_0 = 1.2$, $\rho_2/\rho_0 = 1.4$, as well as the Chaplygin approximation. The functions are replotted on Fig. 2 in the (p, ρ)-plane. It will be noted on Fig. 2 that the slope $a^2 = dp/d\rho$ *increases* with p and ρ along the adiabatic, but *decreases* as ρ increases along the approximating hyperbolic segments; however, at each vertex ρ_i the change of slope a^2 makes up for accumulated divergence between the a^2 values throughout the interval. Fig. 3 shows w^2, a^2, M^2 plotted vs. $1/\rho^2$; the dotted curves give the same quantities for the adiabatic (2.10) (for $\gamma = 1.4$). In the calculations leading to the curves of Figures 1, 2 and the figures that follow, it is assumed that $p_0 = 1$, $\rho_0 = 1$. This is not a serious restriction since by a choice of units any values of p_0, ρ_0 can be reduced to this case. The reader who is irked by this procedure, however, will prefer to replace w^2 by $w^2\rho_0/p_0 = \gamma w^2/a_0^2$, and a^2 by $a^2\rho_0/p_0 = \gamma a^2/a_0^2$.

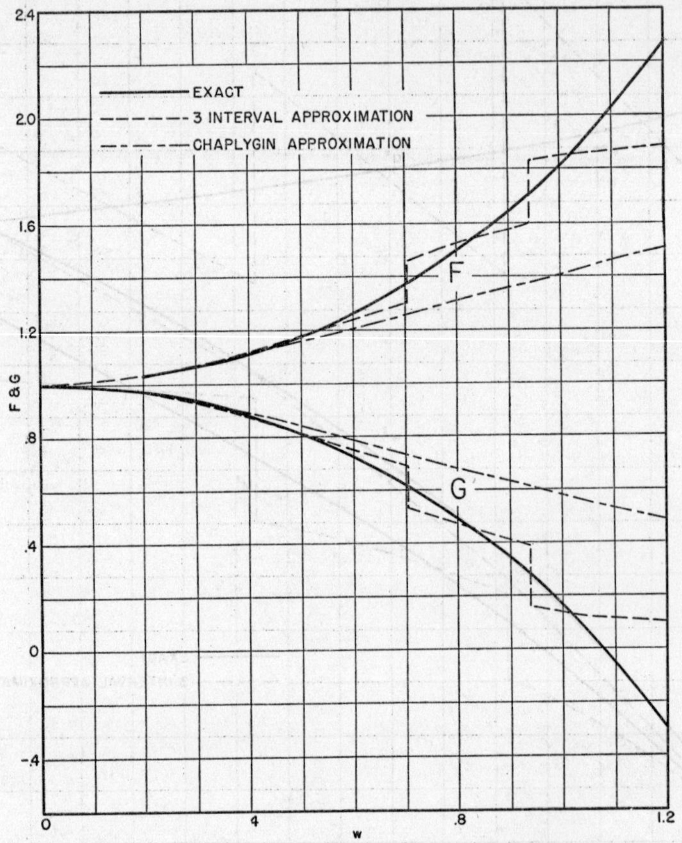

Fig. 4. The coefficients F and G of Chaplygin's differential equation and their approximations

On account of its discontinuities and the gaps in its values, the Mach number M is obviously not a suitable variable to use in the polygonal approximation method.

Elimination of ρ, a from (2.23) under the assumption (3.1) leads to (2.24) but with F, G given not by (2.25) but by

$$(3.9) \quad F(w) = 1 + \frac{w^2}{w^2 + C_i}, \qquad G(w) = \frac{C_i}{w^2 + C_i}, \qquad w_{i-1} < w < w_i .$$

The values of F, G from (2.25) and (3.9) are plotted on Fig. 4 for the example of Figs. 1, 2. Also shown on Fig. 4 are the values of F, G based on the Chaplygin's approximation.

The present method may thus be viewed as consisting in approximating to the coefficients F, G, (2.25) in (2.24) by means of (3.9), this choice of the approximations having the advantage that (2.24) becomes the Laplace equation in W, θ. At w_i the solutions ψ of (2.24) are continuous and so are $\partial\psi/\partial w$.

TABLE I

M	w	$Q_1 = wY_1$	R_3	R_C	$R_3 - Q_1$	$R_C - Q_1$
0	.0000	.0000	0	0	0	0
.1	.11832	.1180	.1177	.1180	−.0003	0
.2	.23575	.2334	.2328	.2335	−.0006	.0001
.3	.35180	.3441	.3428	.3444	−.0013	.0003
.4	.46591	.4481	.4452	.4491	−.0029	.0010
.5	.57736	.5438	.5398	.5466	−.0040	.0028
.55	.63194	.5882	.5840	.5923	−.0042	.0041
.60	.68565	.6300	.6262	.6361	−.0038	.0061
.65	.73854	.6694	.6658	.6779	−.0036	.0085
.70	.79041	.7061	.7019	.7177	−.0042	.0116
.75	.84133	.7403	.7358	.7639	−.0045	.0236
.80	.89127	.7719	.7673	.7916	−.0046	.0197
.85	.94012	.8010	.7963	.8257	−.0047	.0247
.90	.98786	.8275	.8230	.8580	−.0045	.0305
.95	1.03456	.8516	.8471	.8887	−.0045	.0371
1.00	1.08013	.8735	.8692	.9177	−.0043	.0442
1.10	1.16788	.9105	.9069	.9712	−.0036	.0607
1.20	1.25108	.9394	.9385	1.0200	−.0009	.0806
1.30	1.32979	.9612	.9656	1.0620	+.0044	.1008
1.40	1.40402	.9769	.9886	1.1004	.0117	.1235

The discrepancies in F, G on Fig. 4, deriving from the slopes on Fig. 2, are quite appreciable. To study the discrepancies of the solutions, a product type of solution

$$\psi = wY_1(\tau) \cos \theta \tag{3.10}$$

of the exact equations (2.24), (2.25) was compared with a corresponding solution

$$\psi = R \cos \theta, \qquad R = \text{function of } w = R(\tau) \tag{3.11}$$

of (2.24), (3.9), choosing $R = 0$, $dR/dr = 1$ at $r = 0$ so that R agrees with wY in value and slope at $r = 0$. Since $R \cos \theta$ is harmonic in the ζ-plane, it follows that the factor R in (3.11) is a linear combination of r and $1/r$ in each interval (r_i, r_{i+1}). In the first interval the coefficient of $1/r$ is chosen as 0, and in passing from one interval to the next one the new coefficients are determined so as to render R continuous and so that (3.11) satisfies the condition (3.6).

Table I gives the values of wY_1, of R_3, the function R for the three-interval approximation of Fig. 1, and of R_C, the function R for the Chaplygin approximation. Also shown are the differences $R_3 - wY_1$, $R_C - wY_1$. The values Y_1 were taken from Garrick and Kaplan [2]. The Mach number refers to the exact solution. It will be noticed that the discrepancies are far smaller than one would expect on the basis of the differences in F and G. The table was continued into the supersonic interval where the third line segment on Figure 1 gives a rather poor approximation to the adiabatic; the agreement there, while not as good as for $w < w_3$, is surprisingly good. Since the Chaplygin straight line is tangent to the adiabatic at p_0, $1/\rho_0$, the discrepancy close to $w = 0$ (and as far as $w = .6$) is smaller for the one-interval solution, but for larger w-values the Chaplygin solution becomes much worse.

4. **Solutions by the method of reflections.** The product solution method for the determination of ψ was explained in [1] and utilized in Table I. Another method, the method of reflections or images, is based on the following proposition.

Let $h(W, \theta)$ be an arbitrary harmonic function. Then either

$$(4.1)\qquad \begin{aligned} u &= h(W, \theta) + \frac{1 - c_i}{1 + c_i} h(2W_i - W, \theta) \quad \text{in} \quad W > W_i, \\ u &= \frac{2}{1 + c_i} h(W, \theta) \quad \text{in} \quad W < W_i, \end{aligned}$$

or

$$(4.2)\qquad \begin{aligned} u &= h(W, \theta) + \frac{c_i - 1}{c_i + 1} h(2W_i - W, \theta) \quad \text{in} \quad W < W_i, \\ u &= \frac{2c_i}{1 + c_i} h(W, \theta) \quad \text{in} \quad W > W_i, \end{aligned}$$

furnishes a function u satisfying the boundary condition (3.5) at $W = W_i$. Equation (4.1) is useful in finding a function u having the same singularity as h (for instance that of a point source or vortex) in $W > W_i$ and satisfying (3.5), while equation (4.2) furnishes a function u satisfying (3.5) and possessing the same singularities as h in $W < W_i$.

In applying the above to cases with several discontinuities, one runs into the difficulty of obtaining images of images. Thus, consider the case of three-interval approximations with two lines of discontinuity at W_1, W_2, and assume that h has a singularity at the point $W = B$ lying between W_1, W_2. There is no loss in generality in assuming $W_1 = 0$. This may be accomplished by proper choice of K_1 in (2.37). We also put $W_2 = C$.

Application of (4.1) at $W_1 = 0$ shows that (3.5) will be satisfied by the addition to h of

$$(4.3)\qquad h_{-1} = \frac{1 - c_1}{1 + c_1} h(-W, \theta);$$

this term possesses a singularity at $W = B_{-1} = -B_1$, the image of B in $W = 0$, of "strength"

$$R_1 = \frac{1 - c_1}{1 + c_1}. \tag{4.4}$$

Similarly, application of (3.5) with W_2 replaced by C shows that to satisfy (3.5) one may add to h

$$h_1 = \frac{c_2 - 1}{c_2 + 1} h(2C - W, \theta) \tag{4.5}$$

thus placing a singularity at $W = B_1 = 2C - B$, the image of B in $W = C$, of "strength"

$$R_2 = \frac{c_2 - 1}{c_2 + 1}. \tag{4.6}$$

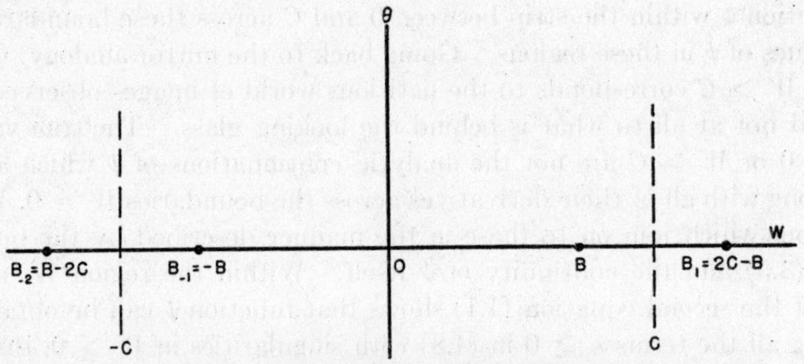

Fig. 5. Reflection of singularity, W, θ-plane

It will be found that the addition of (4.6) now spoils the condition (3.5) over $W = 0$, and it is necessary to add another term, namely

$$h_{-2} = R_1 h_1(-W, \theta) = R_1 R_2 h(W - 2C, \theta) \tag{4.7}$$

to restore this condition; this term is singular at $W = B_{-2} = B - 2C$, the image of B_1 in $W = 0$. Likewise, the addition of h_{-1} as well as of h_{-2} in no way helps the boundary condition (3.5) at $W = C$, and additional terms are required to restore this condition. This process can clearly be continued leading to further singularities over a periodic array of points B_n which are the successive images of B in the two rectilinear boundaries, $W = 0$, $W = C$ (See Fig. 5). The position of these points is analogous to what one observes in a room with two parallel mirrors on opposite walls. Upon looking into either one, one sees one's own successive reflections extending to infinity and obtained by reflections back and forth across each mirror. Moreover, the intensity of each reflection is obtained from the image which is being reflected by multiplication by a proper "coefficient of reflection," which in the present case is equal to R_1 for reflection over the left-hand mirror $W = 0$, and R_2 for reflection across the right-hand mirror. There results the following infinite series

$$\psi = \sum_{n=-\infty}^{+\infty} h_n(W, \theta) \qquad \text{for } 0 < W < C \tag{4.8}$$

where

$$\begin{aligned} h_0(W, \theta &= h(W, \theta), \\ h_{-n}(W, \theta) &= R_1 h_{n-1}(-W, \theta), \qquad n > 0, \\ h_n(W, \theta) &= R_2 h_{1-n}(2C - W, \theta), \qquad n < 0. \end{aligned} \tag{4.9}$$

The above infinite series converges provided that both R_1 and R_2 are numerically less than 1 and h does not become infinite "too rapidly" at infinity. The convergence is then as rapid as that of a geometric series.

It must be clearly understood that the series (4.8) applies *only* in the strip $0 < W < C$ in Fig. 5 within which the singularity B lies. While the series (4.8) converges outside of that strip too, it represents the *analytic continuation* of the function ψ within the strip between 0 and C across these boundaries, and *not* the values of ψ in these regions. Going back to the mirror analogy, (4.8) for $W < 0$ or $W > C$ corresponds to the fictitious world of images observed in the mirror, and not at all to what is behind the looking glass. The true values of ψ in $W < 0$ or $W > C$ are not the analytic continuations of ψ which are continuous along with all of their derivatives across the boundaries $W = 0$, $W = C$, but functions which join on to these in the manner described by the boundary condition (3.5) and the continuity of ψ itself. Within the region $W < 0$ application of the second equation (4.1) shows that function ψ can be obtained by multiplying all the terms $n \geqq 0$ in (4.8) with singularities in $W > 0$, that is, at $B, B_1, B_2 \cdots$, by the factor $1 + R_1$. In this way there results the infinite series

$$\psi = \frac{2}{1 + c_1} \sum_{n=0}^{\infty} h_n(W, \theta) \qquad \text{for } W < 0. \tag{4.10}$$

Similarly, one obtains from (4.2), (4.8)

$$\psi = \frac{2c_2}{1 + c_2} \sum_{n=0}^{-\infty} h_n(W, \theta) \qquad \text{for } W > C. \tag{4.11}$$

Thus ψ is actually free from singularities except at B.

Introducing ζ as in (2.38), one replaces the lines $W = 0$, $W = C$ by the circles $r = r_1 = 1$, $r = r_2 = e^C$, while the singular points B_n are transformed into

$$b_n = e^{B_n} \tag{4.12}$$

so that

$$\cdots, \quad b_{-1} = 1/b_0, \quad b_0 = b = e^B, \quad b_1 = e^{2C}/b_0, \quad b_2 = be^{2C}, \cdots. \tag{4.13}$$

The reflections across $W = 0$, $W = C$ now correspond to inversions across the circles $r = 1$, $r = r_2 = e^C$. An additional singularity at $\zeta = 0$ corresponding to $W = -\infty$ may now appear, unless special provisions are made to eliminate it.

In practice it may be convenient to use the series (4.10) or (4.11) for computing ψ in the range $0 < W < C$. This is done for (4.11) by identifying ψ in (4.11) with u in the second equation (4.2), solving for h and substituting in the first equation (4.2).

The nature of equations (4.8)–(4.11) can be clarified by means of Fig. 6 which is based on the optical analogy mentioned above, but with the mirrors made "semi-transparent." The terms in the series (4.8)–(4.11) are represented by rays which start from the point B of Fig. 6 at a fixed angle with the horizontal

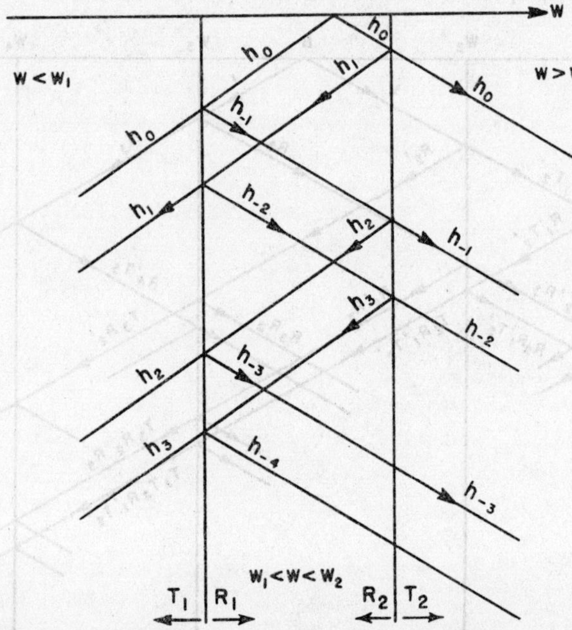

Fig. 6. Reflection ray analogy for finding images of singularities for the case of three rectilinear intervals

and are reflected and transmitted at the boundaries W_1 and W_2 with reflection coefficients R_1, R_2 given by (4.4), (4.5) and transmission coefficients

$$T_1 = 1 + R_1, \qquad T_2 = 1 + R_2. \tag{4.14}$$

The transmitted rays T_1h_1, T_1h_2, $\cdots$ can be visualized as originating from the source points $B_1, B_2 \cdots$ on the W-axis, the transmitted rays T_2h_{-1}, T_2h_{-2}, $\cdots$ as originating from the source points B_{-1}, B_{-2}, $\cdots$.

The optical analogy method is equally applicable to the case of more than three intervals. The case of five intervals is represented in Fig. 7. One now introduces reflection and transmission coefficients R_i, T_i for rays striking W_i from the left, R_i', T_i' for rays striking W_i from the right, where

$$\begin{aligned} R_i' &= \frac{1 - c_i}{1 + c_i}, \qquad T_i' = \frac{2}{1 + c_i} = 1 + R_i', \\ R_i &= \frac{c_i - 1}{c_i + 1} = -R_i', \qquad T_i = \frac{2c_i}{c_i + 1} = 1 + R_i, \end{aligned} \tag{4.15}$$

c_i being defined by (3.5). Corresponding to each ray segment there is a term obtained by multiplying the function

$$h(\quad, \theta) \tag{4.16}$$

by the coefficient placed on the ray, where the first argument is obtained by replacing W by $2W_i - W$ corresponding to *each* reflection the ray suffered on its way from B. (Transmissions do not affect the first argument in 4.16.)

The increased complexity due to the multiple boundaries as the number of

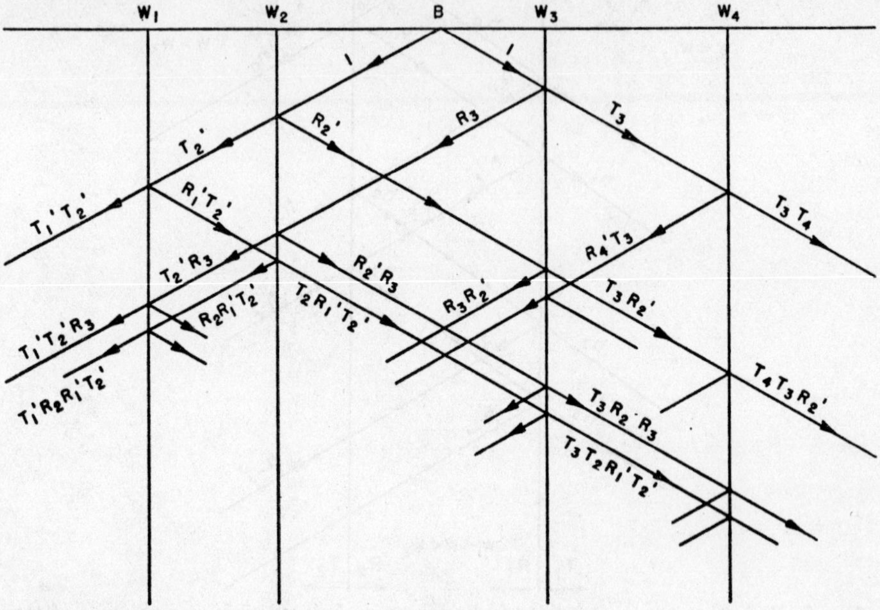

Fig. 7. Reflection ray analogy for finding images of singularities for the case of five rectilinear intervals

intervals increases is partly compensated for by the decreased magnitudes of R_i, R_i'.

It is to be kept in mind that even in relatively simple flows in the physical plane, the function ψ may be multiple-valued in the hodograph plane. Since in the solutions indicated above h is not restricted to single-valued functions, these solutions are applicable even to the multiple-valued cases.

5. **Example.** We apply the method of the preceding paragraph to the flow function ψ which corresponds to a point source at $\zeta = b$, and put

$$h = \text{Im}\{\log(\zeta - b_0)\} = \theta_0, \tag{5.1}$$

where θ_0 is the argument of $\zeta - b_0 = \zeta - b$, that is, the angle between this complex vector and the real ζ-axis. Inversion of $f(\zeta)$ in the circle $r = R$ is accomplished by replacing ζ by $R^2/\bar{\zeta}$ leading to $f(R^2/\bar{\zeta})$, bars denoting conjugates. For the case $f(\zeta) = \log(\zeta - b)$, this leads to

$$\log(R^2/\bar{\zeta} - b) = \log(\bar{\zeta} - R^2/b) - \log\bar{\zeta} + \log(-b). \tag{5.2}$$

With $R = 1$, $R^2/b = 1/b = b_{-1}$, the imaginary part of (5.2) yields (see Fig. 8)

(5.3) $$-\theta_{-1} + \theta \pm \pi, \qquad \theta_{-1} = \arg(\zeta - b_{-1}), \qquad \theta = \arg \zeta.$$

This leads to

(5.4) $$h_{-1} = R_1(-\theta_{-1} + \theta) + \text{const.}$$

Similar calculations of h_n for other n yield

(5.5) $$\psi = \sum_{n=-\infty}^{+\infty} e_n \theta_n - \theta \sum_{n=-1}^{-\infty} e_n - \pi \sum_{n=1}^{+\infty} e_n \qquad \text{for } r_1 < r < r_2,$$

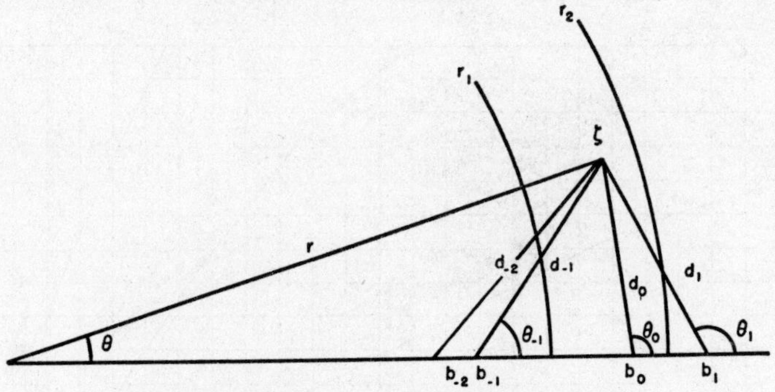

Fig. 8. Reflection of singularity, r, θ-plane

where

$$\theta_n = \arg(\zeta - b_n) = \text{Im} \log(\zeta - b_n),$$

(5.6) $$e_{-1} = -R_1, e_1 = -R_2$$

$$e_{\pm 2} = R_1 R_2, \qquad e_{\pm 3} = e_{\mp 2} e_{\pm 1} = -R_1 R_2 R_1^2, \cdots.$$

Inside the (unit) circle $r = r_1$ and outside the circle $r = r_2$ one obtains similarly from (4.10), (4.11)

(5.7) $$\psi = \pi + \frac{2}{1 + c_1} \sum_{n=0}^{\infty} e_n(\theta_n - \pi) \qquad \text{for } r < r_1,$$

(5.8) $$\psi = \frac{2c_2}{1 + c_2} \left(\sum_{n=0}^{-\infty} e_n \theta_n - \theta \sum_{n=0}^{-\infty} e_n \right) + \theta \qquad \text{for } r > r_2.$$

In choosing the θ-terms and the constant terms in (5.5), (5.7), (5.8), a slight departure from direct inversion in accordance with (4.8), (5.2) has been made so that without violating (3.6) the function ψ is made free from singularities at $r = 0$ and vanishes for real $\zeta > b$.

As an example, the flow function ψ was calculated for the case of a point source at $b = b_0 = 1.15$ for the (p, ρ)-approximations described in §3 and shown on Figs. 1, 2. The images of the point source occur at

$$b_1 = 1.2861, \quad b_2 = 1.7008, \quad b_3 = 1.9021, \quad b_4 = 2.5155, \cdots,$$

$$b_{-1} = .8696, \quad b_{-2} = .7776, \quad b_{-3} = .5880, \quad b_{-4} = .5257, \cdots.$$

The coefficients c_1, c_2 are given by

$$c_1 = 1.1388, \quad c_2 = 1.5633$$

leading to

$$e_{-1} = -R_1 = .06490, \qquad e_1 = -R_2 = -.21975$$

and to

$$e_{-2} = e_2 = -.01426, \quad e_{-3} = -.00093, \quad e_3 = .00313, \quad e_{-4} = e_4 = .00020, \cdots.$$

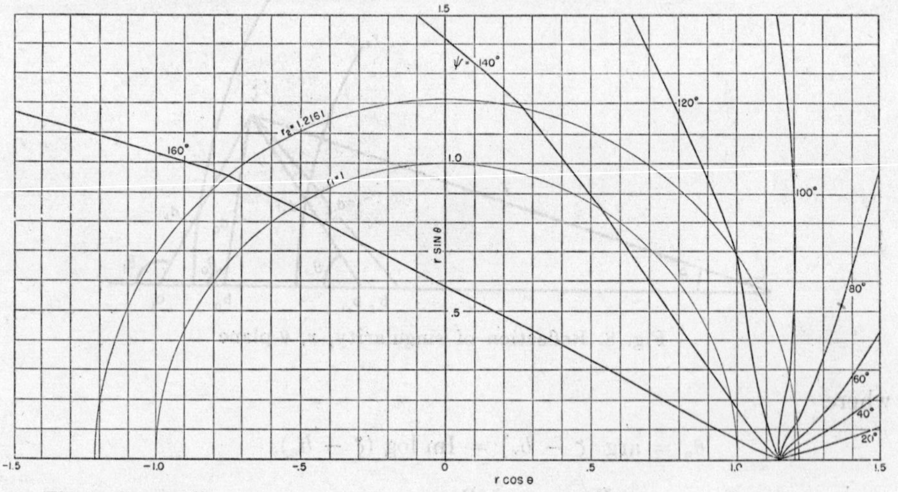

Fig. 9. Stream lines ψ = const. in the ζ-plane for point source at $r = 1.15$, $\theta = 0$

The series (5.5), (5.7) and (5.8) become, with θ_i, ψ expressed in degrees ($\Delta\psi = 360°$ for unit point source):

$$\psi = 180° + .9351(\theta_0 - 180°) - .2055(\theta_1 - 180°) - .0133(\theta_2 - 180°)$$
$$+ .0029(\theta_3 - 180°) + .0002(\theta_4 - 180°) + \cdots, \qquad \text{for } 0 \leqq r \leqq r_1,$$

$$\psi = 41.522^0 - .0499\theta + \theta_0 + .2198\theta_1 - .0143\theta_2 + .0031\theta_3 + .0002\theta_4 + \cdots$$
$$+ .0694\theta_{-1} - .0143\theta_{-2} - .0009\theta_{-3} + .0002\theta_{-4} + \cdots, \qquad r_1 \leqq r \leqq r_2,$$

$$\psi = -.2806\theta + 1.2198\theta_0 + .0792\theta_{-1} - .0174\theta_{-2} - .0011\theta_{-3} + .0002\theta_{-4} + \cdots, r \geqq r_2.$$

The plot of the curves ψ = constant in the $\zeta = re^{i\theta}$ plane is shown in Fig. 9 for $\psi = 0°, 20°, 40°, \cdots$. Figure 10 shows the same flow lines but in the hodograph plane.

As a further example, the field of a point vortex was considered. It will be recalled from equation (2.38) that the conjugate harmonic of ψ is not a possible flux function (since $-(B/C)^{1/2}\varphi + i\psi$ is analytic in ζ and not $\varphi + i\psi$). Hence, unlike the incompressible case, the conjugate harmonic of the point source solution will not do for a point vortex. Going back to (4.8)–(4.11) we put

(5.9) $$h_0 = \mathrm{Re}[\log(\zeta - b)] = \log d_0, \qquad d_0 = |\zeta - b|$$

(this identifies $f(\zeta)$ in (2.38) with $i \log(\zeta - b)$). Inversion in $r = R$ yields (see Fig. 8)

(5.10) $$\log(\zeta - b) \to \log(R^2/\bar{\zeta} - b) = \log(\bar{\zeta} - (R^2/b)) - \log \bar{\zeta} + \log(-b).$$

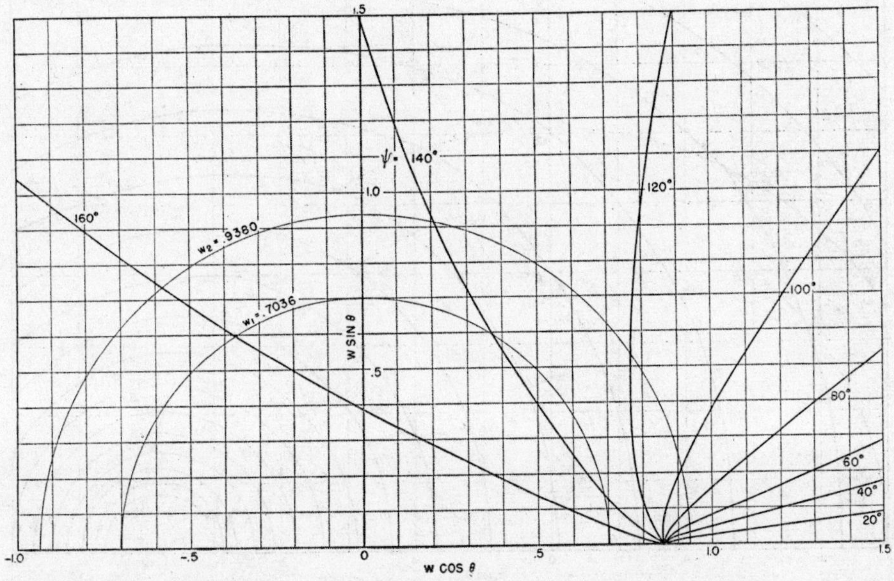

Fig. 10. Stream lines ψ = const. in the hodograph plane for point source at $w = .8599$, $\theta = 0$

On applying to $R = 1$ and taking real parts one obtains

(5.11) $$\log d_{-1} - \log r + \log b, \qquad d_{-1} = \zeta - b_{-1}.$$

Comparison with (5.3) and (5.4) shows that

(5.12) $$h_{-1} = R_1(\log d_{-1} - \log r) + \text{const.}$$

Application to (4.9)–(4.11) with $h_0 = \log r_0$ leads to

(5.13) $$\psi = \frac{2}{1 + c_1} \sum_{n=0}^{+\infty} (-1)^n e_n \log d_n, \qquad \text{for } r \leq r_1 = 1,$$

(5.14) $$\psi = \sum_{n=-\infty}^{+\infty} (-1)^n e_n \log d_n - \log r \sum_{n=-1}^{-\infty} e_n(-1)^n + K_1, \quad \text{for } r_1 \leq r \leq r_2,$$

$$(5.15)\qquad \psi = \frac{2c_2}{1+c_2}\left[\sum_{n=0}^{\infty}(-1)^n e_n \log d_n\right] + K_2 \log r + K_3, \qquad \text{for } r \geqq r_2,$$

where the constants K_1, K_2, K_3 are so determined that ψ is continuous at r_1 and r_2, and (3.6) holds. The latter condition is best applied by means of

$$(5.16)\qquad \int \frac{\partial\psi}{\partial r}\bigg|_{r_i+} ds = c_i \int \frac{\partial\psi}{\partial r}\bigg|_{r_i-} ds$$

leading to an equation involving the sum of the coefficients of the logarithms whose argument vanishes inside the circle of integration.

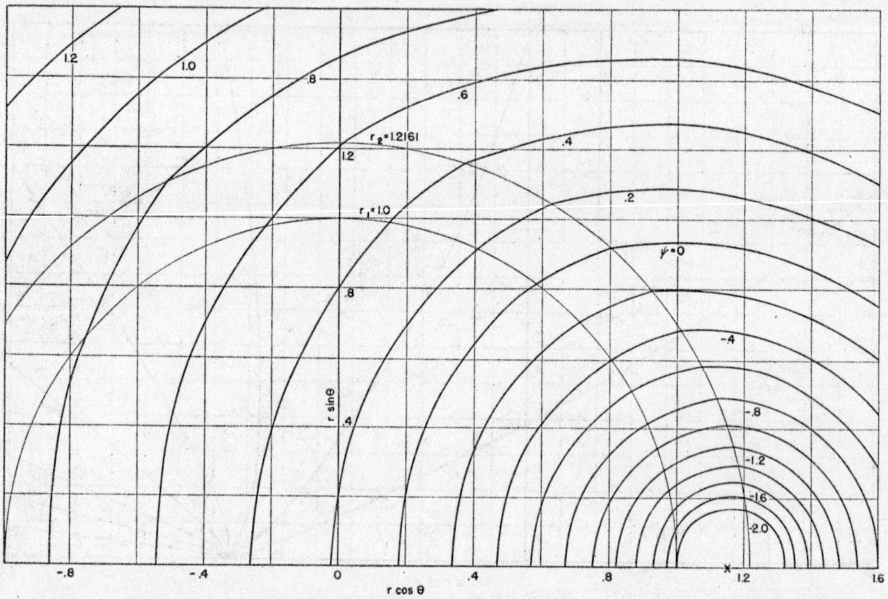

Fig. 11. Stream lines ψ = const. in the ζ-plane for point vortex at $r = 1.15$, $\theta = 0$

Substitution of the previous values of r_1, r_2, b, e_n, b_n, c_1, c_2 into (5.13)–(5.15) leads to the following three series

$$\psi = .9351 \log d_0 + .2055 \log d_1 - .0133 \log d_2 - .0029 \log d_3 + .0002 \log d_4 + \cdots, \qquad \text{for } r \leqq r_1,$$

$$\psi = -.0120 + .0890 \log r + \log d_0 + .2198 \log d_1 - .0143 \log d_2 - .0031 \log d_3 + .0002 \log d_4 + \cdots - .0649 \log d_{-1} - .0143 \log d_{-2} + .0009 \log d_{-3} + .0002 \log d_{-4} + \cdots, \qquad \text{for } r_1 \leqq r \leqq r_2,$$

$$\psi = -.0764 + .4387 \log r + 1.2198 \log d_0 - .0792 \log d_{-1} - .0174 \log d_{-2} + .0011 \log d_{-3} + .0002 \log d_{-4} + \cdots, \qquad \text{for } r \geqq r_2.$$

The flow lines for the above ψ are plotted on Fig. 11 for $\psi = 0, \pm.2, \pm.4, \cdots$. On Fig. 12 the aspect of the same flow lines in the hodograph plane is given.

It must be kept in mind that the examples just discussed, while they possess the required point-source singularity, are *not unique* solutions. It is clear that any solution which is free from singularities in the regions under consideration may be added to the above.

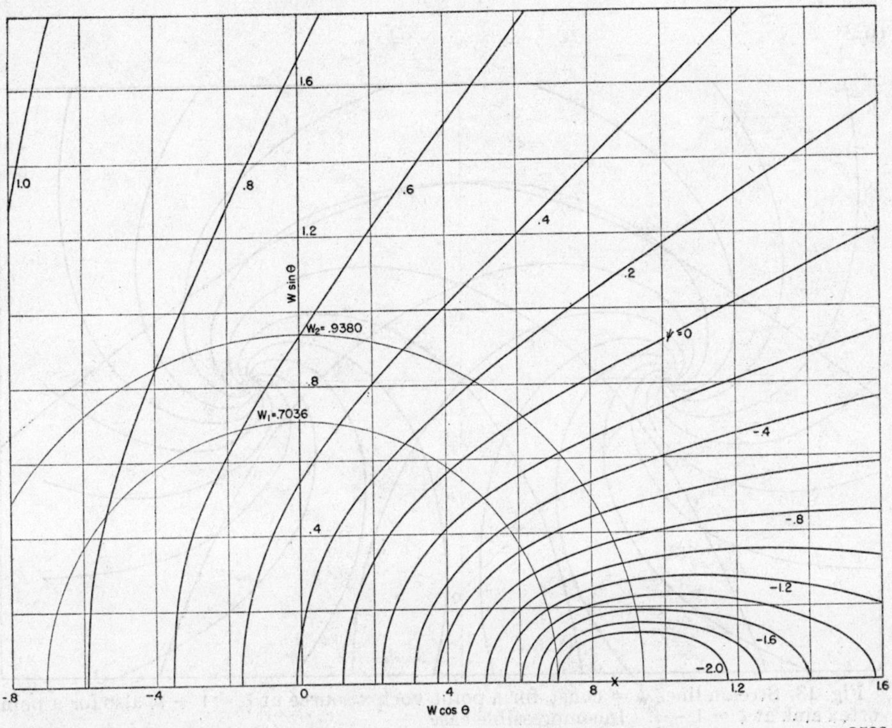

Fig. 12. Stream lines ψ = const. in the hodograph plane for point vortex at $w = .8599$, $\theta = 0$

It is believed that the method outlined above is far more convenient from computational point of view than other methods that have been developed for handling these flows, for instance the one given in [3].

6. **Determination of the flow in the physical plane.** The point source and the point vortex solutions are of interest in connection with the flow through a grid of similar blades, the entering flow at infinity corresponding to a point source and a point vortex; the exit flow at infinity corresponding to a point sink and vortex. It is clear, however, that further functions ψ would have to be added to the above solutions to obtain blade shapes of practical interest. Nevertheless, it is of interest to find the aspect of the physical flow arising from a pure point vortex-source and a point vortex-sink. This can be determined by superposition of the above solutions and carrying out the integrations (2.40). For

the present, however, this calculation was carried out only for the incompressible case for which the above equations simplify considerably.

For the incompressible case, the variable ζ may be directly identified with the hodograph variable

$$\zeta = we^{i\theta} = u + iv. \tag{6.1}$$

From (2.35) it follows that the complex potential

$$\omega = \varphi - i\psi/\rho_0 \tag{6.2}$$

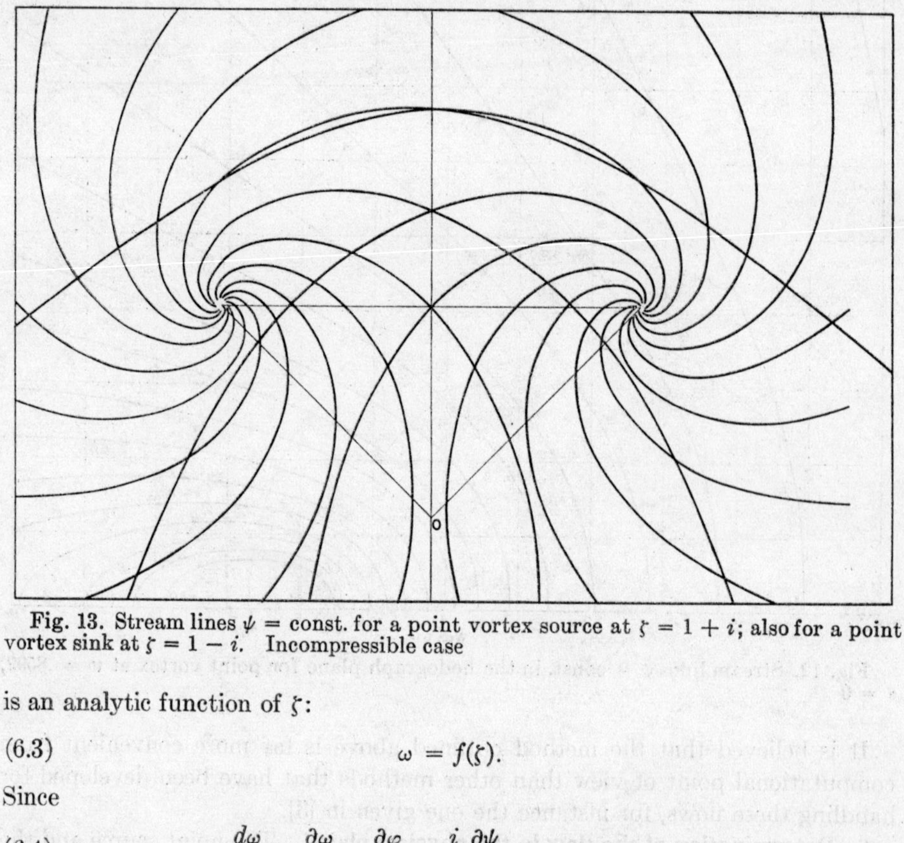

Fig. 13. Stream lines ψ = const. for a point vortex source at $\zeta = 1 + i$; also for a point vortex sink at $\zeta = 1 - i$. Incompressible case

is an analytic function of ζ:

$$\omega = f(\zeta). \tag{6.3}$$

Since

$$\frac{d\omega}{dz} = \frac{\partial \omega}{\partial x} = \frac{\partial \varphi}{\partial x} - \frac{i}{\rho_0}\frac{\partial \psi}{\partial x} = u + iv = \zeta \tag{6.4}$$

we have

$$z = \int dz = \int \frac{d\omega}{\zeta}. \tag{6.5}$$

Placing the point source and the point vortex, each of unit strength, at the point $\zeta = 1 + i$ and the point sink and the point vortex of similar strength at the point $\zeta = 1 - i$, we have

$$\omega = A \log (\zeta - a) + B \log (\zeta - b), \tag{6.6}$$
$$A = 1 + i, \quad B = -1 + i, \quad a = 1 + i, \quad b = 1 - i.$$

The integration (6.5) yields

$$\begin{aligned} z &= A \int \frac{d\zeta}{\zeta(\zeta - a)} + B \int \frac{d\zeta}{\zeta(\zeta - b)} \\ &= \frac{A}{a} [-\log \zeta + \log (\zeta - a)] + \frac{B}{b} [-\log \zeta + \log (\zeta - b)]. \end{aligned} \tag{6.7}$$

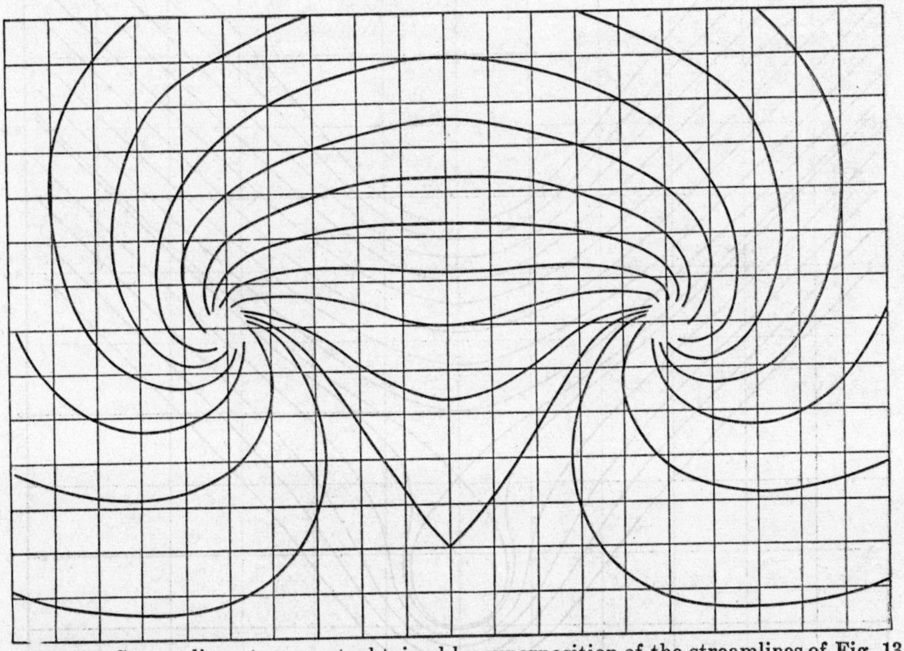

Fig. 14. Stream lines ψ = const. obtained by superposition of the streamlines of Fig. 13 Incompressible case

For the values indicated in (6.6) this reduces to

$$\omega = (1 + i) \log (\zeta - a) + (-1 + i) \log (\zeta - b), \tag{6.8}$$

$$z = \log (\zeta - a) - \log (\zeta - b). \tag{6.9}$$

The lines of flow in the hodograph were first obtained by superposition of the rectilinear flow lines corresponding to the point source and the circular flow lines corresponding to the point vortex leading to equiangular spirals passing through diagonally opposite corners of the resulting small squares formed by the above radial lines and circles. After the above logarithmic spirals have been constructed around the point $\zeta = 1 + i$, a similar set was obtained for the $\zeta = 1 - i$ as indicated in Fig. 13, and by superposition the streamlines of Fig. 14 were obtained. Finally, by substituting the values of ζ for these flow lines into (6.9) the flow lines in the physical plane were found. These are shown on Fig. 15.

It will be noted from equation (6.9) that the integration leading from the hodograph to the physical plane possesses no singularity at the origin $\zeta = 0$ as might ordinarily be expected from (6.5). The resulting flow is smooth and analytic in the whole physical plane, and corresponds to a 90° turn of the flow from its incident direction at infinity to its exit direction at infinity.

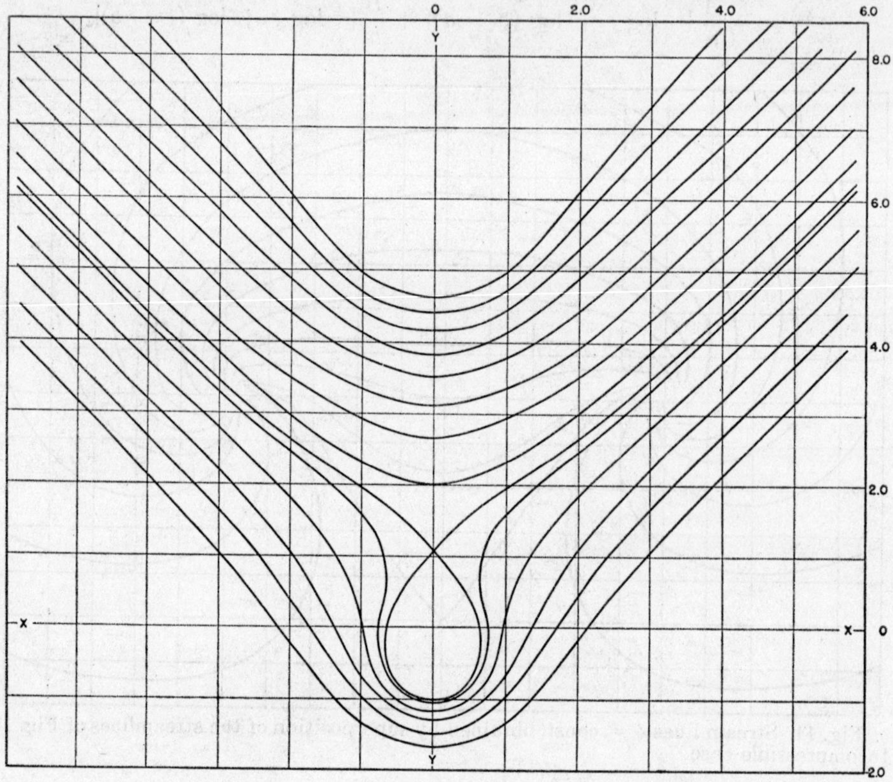

Fig. 15. Stream lines ψ = const. in physical plane corresponding to Fig. 14

BIBLIOGRAPHY

1. H. Poritsky, *An approximate method of integrating the equations of compressible fluid flow in the hodograph plane*, Proceedings of the Sixth International Congress of Applied Mechanics, Paris, 1946.

2. I. E. Garrick and Carl Kaplan, I. *On the flow of a compressible fluid by the hodograph method*. II. *Fundamental set of particular flow solutions of the Chaplygin differential equaion*, NACA ARR, No. L4I29.

3. H. S. Tsien and Y. H. Kuo, *Two-dimensional irrotational mixed subsonic and supersonic flow of a compressible fluid and the upper critical Mach number*, NACA Technical Note 995.

GENERAL ELECTRIC COMPANY,
SCHENECTADY, N.Y.

THE BOUNDARY LAYER OF YAWED CYLINDERS

BY

W. R. SEARS

Starting with the Navier-Stokes equations of motion for a viscous fluid, we repeat Prandtl's argument regarding the orders of magnitude of quantities in a laminar boundary layer, for the special case of a yawed cylinder of infinite length. The equations resulting are as follows:

$$u_t + uu_x + wu_z = -p_x/\rho + \nu u_{zz}, \tag{1}$$

$$v_t + uv_x + wv_z = \nu v_{zz}, \tag{2}$$

$$0 = p_z, \tag{3}$$

$$u_x + w_z = 0, \tag{4}$$

where the coordinates x and z are measured along and perpendicular to the surface, respectively, and y is measured in the direction of the cylindrical axis; u, w, and v are the corresponding velocity components; p is the pressure; ρ the density (assumed constant); t the time and ν the kinematic viscosity. Since the cylinder is uniform and infinite in extent in the y direction, it follows that u, v, and w are functions of x, z, and t only. The pressure $p(x, t)$ is supposed to be given by the potential-flow conditions at the outer edge of the boundary layer.

It is seen that (1), (3), and (4) are the customary non-linear equations describing the two-dimensional boundary-layer flow about an un-yawed cylinder, while (2) constitutes a linear equation by which $v(x, z, t)$ can be computed if the two-dimensional problem is solved. Moreover, the two-dimensional problem involved is exactly the case of plane flow about the same cylinder in its un-yawed condition.

It may be useful to notice that, in the steady-flow case, equation (2) is the same as the equation for the temperature distribution $T(x, z)$ in the laminar boundary layer of the un-yawed cylinder, provided that the Prandtl number is one and viscous heating is neglected. Thus the distribution of spanwise velocity near a yawed cylinder is the same as the distribution of temperature near the same cylinder, at zero yaw, when its surface is maintained at a uniform temperature different from the stream temperature.

The separation of the equations achieved here leads to interesting conclusions about certain boundary-layer phenomena on a yawed infinite cylinder. For example, the chordwise location of laminar separation (of the chordwise flow), which is determined by equations (1), (3), and (4), must be invariant as the cylinder is yawed. In addition, it may be speculated that the phenomenon of transition is controlled by the Reynolds number based on the chordwise component of the stream velocity.

The first special case treated is that of a yawed flat plate in steady flow. Here

p_x is identically zero, and it is easily seen that u and v are proportional. In fact, the result means that the boundary-layer flow is always in the free-stream direction with the so-called Blasius velocity profile, that is, is unaffected by yaw of the leading edge.

We proceed next to consider yawed steady flow about a class of cylinders suggested by Prandtl, for which the potential velocity outside the boundary layer can be expressed in the form

$$u_1(x) = a_1x + a_3x^3,$$

a_1 and a_3 being constants. The plane problem has been solved, and the functions defining $u(x, z)$ and $w(x, z)$ are available in tabular form;[1] for example

$$u = a_1xf_1'(\eta) + a_3x^3f_3'(\eta) + \cdots$$

where $\eta \equiv (UL/\nu)^{1/2} z/L$, U being the chordwise component of the stream velocity and L a characteristic dimension. To obtain the appropriate solution of (2) in this case, we assume

$$v = V\sum_{n=0}^{\infty} (x/L)^n G_n(\eta)$$

where V is the axial component of the stream velocity. This leads to a system of ordinary differential equations for the $G_n(\eta)$. The boundary conditions require $G_1 = G_3 = \cdots = 0$. The even-numbered functions have been evaluated numerically for $n = 0, 2, 4$, and 6. It appears that these are sufficient for an estimate of the spanwise flow in the boundary layer, forward of separation.

The results include diagrams showing the magnitude and direction of boundary-layer flow at various positions forward of the line of separation of chordwise flow. The shape of the "limiting streamline" at the surface of the cylinder has also been determined and is drawn in comparison with the S-shaped streamline of the exterior potential flow.

CORNELL UNIVERSITY,
ITHACA, N. Y.

[1] S. Goldstein (editor), *Modern developments in fluid dynamics*, Oxford, 1938, pp. 148–152.

ON SHOCK-WAVE PHENOMENA: INTERACTION OF SHOCK WAVES IN GASES

BY

H. POLACHEK AND R. J. SEEGER

1. **Introduction.** Interest in shock-wave phenomena was revived during World War II. The authors were introduced to the subject by J. von Neumann when all were associated with the Navy Bureau of Ordnance. Under his inspiring leadership there was initiated theoretical and experimental research which resulted not only in actual practical success, but also in real intellectual stimulation. Much of the present understanding of shock-wave phenomena is due to his physical insights. It is not inappropriate, therefore, to review his general investigations [13] together with detailed ones of the authors [16] at this first annual postwar symposium in applied mathematics.

The formation and propagation of a shock wave in a compressible substance (non-viscous and non-thermal-conductive) have been matters of discussion ever since the early investigations of W. J. M. Rankine and H. Hugoniot [11]. Primary contributors have been J. W. Rayleigh [17], G. I. Taylor[1] [24, 8], R. Becker [1], and L. H. Thomas [24]. The one-dimensional problem of shock-wave interaction, however, has attracted attention only since 1941 [6]. Von Neumann's work revealed two-dimensional interaction of shock waves to be of even greater interest both physically and mathematically.

The oblique reflection of a shock wave is equivalent to the collision of two similar shocks. It might be supposed that the ordinary scheme of acoustic reflection would be applicable to shock-wave reflection. According to von Neumann [13], however, E. Teller first pointed out to him that there is a breakdown in the case of certain shocks—a breakdown which is associated with that of supersonic flow into a concave corner and with the detachment of the headwave of a wedge in supersonic flow. Von Neumann related this breakdown to the soot *V*-effect of E. Mach.[2] He predicted a new type of reflection, which has subsequently been termed "Mach reflection".

Experimental verification started in 1942 with the repetition of Mach's investigations by E. Bright Wilson, Jr., et al., and later with their extension by R. W. Wood. Direct confirmation was achieved through ballistic spark photography by A. C. Charters and R. N. Thomas in 1942. Valuable contributions were made in 1943 also by P. Libessart and by H. Lean. A systematic quantitative investigation using a so-called shock tube was completed by L. G. Smith [22] in 1944. More precise measurements are now being sought independently by L. G. Smith, W. Bleakney [10], et al. Evaluation of the experimental material

[1] Cf. also G. I. Taylor, *Pressure on solid bodies near an explosion*, RC118.

[2] E. Mach, Vienna Academy, Sitzungsberichte vol. 78 (1878) p. 819; other papers vols. 72–92 (1875–1889).

has been made from time to time by von Neumann [13], P. C. Keenan [12], and the authors [15]. The present summary is concerned not with this interesting experimental history, but rather with the mathematical investigations which have been essential for guiding these experiments.

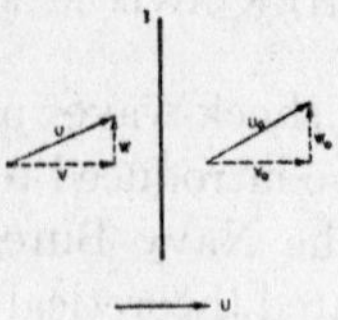

FIG. 1

2. **Plane shock waves in gases.** Consider a simple model [16] of a shock wave, say, a plane step shock wave I (Fig. 1) moving with constant velocity U in a gas of negligible viscosity and thermal conductivity. Let the physical state of the gas initially be specified by its pressure p_0, density ρ_0, and material velocity u_0, and that behind the shock be specified by its pressure p, density ρ, and material velocity u. It is convenient to resolve the material velocity u into components V and W which are normal and tangential to the shock front, respectively; likewise, u_0 may be resolved into components V_0 and W_0. Now continuity of tangential velocity requires the equality of W and W_0, so that we shall not concern ourselves further about these components.

If we consider the flow of the gas relative to the shock front, we have by the conservation of mass

$$(1) \qquad \rho_0(U - V_0) = \rho(U - V).$$

The conservation of momentum requires that

$$(2) \qquad p - p_0 = \rho_0(U - V_0)(V - V_0).$$

Eliminating first V and then V_0 from these two equations, we obtain, respectively,

$$(3) \qquad U - V_0 = \left(\frac{p - p_0}{(\rho_0/\rho)(\rho - \rho_0)}\right)^{1/2}, \qquad U - V = \left(\frac{p - p_0}{(\rho/\rho_0)(\rho - \rho_0)}\right)^{1/2}.$$

On the other hand, the elimination of U from the same equations gives

$$(4) \qquad V - V_0 = \left(\frac{(p - p_0)(\rho - \rho_0)}{\rho\rho_0}\right)^{1/2}.$$

In terms of E, the intrinsic energy per unit mass, the conservation of energy states

$$(5) \qquad E - E_0 = \frac{(p + p_0)(\rho - \rho_0)}{2\rho\rho_0}.$$

It is convenient to introduce a temperature-dependent parameter. One method of defining such a parameter e is by the expression

$$E = (e - 1)p/\rho.$$

Substitution of this value of E in the energy equation (5) yields

$$\frac{\rho_0}{\rho} = \frac{(2e_0 - 1)p_0 + p}{p_0 + (2e - 1)p}.$$

Or, in terms of the density ratio $\eta \equiv \rho_0/\rho$ and of the compression ratio $\xi \equiv p_0/p$

(5a)
$$\eta = \frac{(2e_0 - 1)\xi + 1}{\xi + (2e - 1)}.$$

A second mode of definition is the use of an average value $\bar{e}$ specified by

$$E - E_0 = (\bar{e} - 1)\left(\frac{p}{\rho} - \frac{p_0}{\rho_0}\right).$$

In terms of $\bar{e}$ equation (5) becomes

$$\frac{\rho_0}{\rho} = \frac{(2\bar{e} - 1)p_0 + p}{p_0 + (2\bar{e} - 1)p} \quad \text{or} \quad \eta = \frac{(2\bar{e} - 1)\xi + 1}{\xi + (2\bar{e} - 1)}.$$

In this case a more useful parameter is $\bar{\gamma}$ given by

$$\bar{\gamma} \equiv \bar{e}/(\bar{e} - 1).$$

We now obtain

(5b)
$$\eta = \frac{(\bar{\gamma} + 1)\xi + (\bar{\gamma} - 1)}{(\bar{\gamma} - 1)\xi + (\bar{\gamma} + 1)}.$$

If $\bar{\gamma}$ is constant, it is simply the ratio of specific heats, which is equal to 5/3 for monatomic ideal gases, 7/5 for diatomic ideal gases, and so on. In general, however, $\bar{\gamma}$ is not constant, but represents a mean ratio of specific heats; it depends at high temperatures on molecular vibration, dissociation, and electronic excitation (to a less degree). For temperatures below 2000° K., for which dissociation is not important, the variation of $\bar{\gamma}$ with temperature is given by Becker's formula [1] or by Smallwood's expression [21]. Tables [3, 2, 4, 14] available for states of air at higher temperatures may be used to construct the adiabatic Rankine-Hugoniot curves [16] for the variation of the pressure with the density. By comparison of these more accurate curves with the corresponding curve for $\bar{\gamma} = 1.40$ it is found that the simple assumption of constant $\bar{\gamma}$ is valid for shocks of strength up to $\xi \sim 0.3$. Inasmuch as most of the experimental investigations on the reflection of shocks so far have been within this range, the equations that follow will be given for constant γ (we shall drop the bar).

It is possible to express all the previous relations in terms of a single parameter. The choice of this parameter depends on the particular needs, which vary. We shall choose ξ. Furthermore, it is convenient to express all speeds as ratios with respect to the speed of sound. Let $c_0 \equiv (\gamma p_0/\rho_0)^{1/2}$ be the speed of sound in front of the shock and $c \equiv (\gamma p/\rho)^{1/2}$ be that behind the shock. Then equations (3) can be expressed variously by

$$M \equiv \frac{U - V_0}{c_0} = \left(\frac{(\gamma + 1)/\xi + (\gamma - 1)}{2\gamma}\right)^{1/2},$$

$$(3') \qquad \tau \equiv \frac{U - V_0}{c} = \frac{(\gamma - 1)\xi + (\gamma + 1)}{(2\gamma[(\gamma + 1)\xi + (\gamma - 1)])^{1/2}},$$

$$\sigma \equiv \frac{U - V}{c} = \left(\frac{(\gamma + 1)\xi + (\gamma - 1)}{2\gamma}\right)^{1/2},$$

and equation (4) becomes

$$(4') \qquad -\nu \equiv \frac{V - V_0}{c} = \frac{2(1 - \xi)}{(2\gamma[(\gamma + 1)\xi + (\gamma - 1)])^{1/2}}.$$

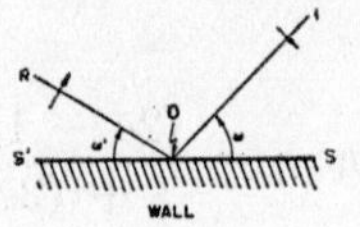

FIG. 2a

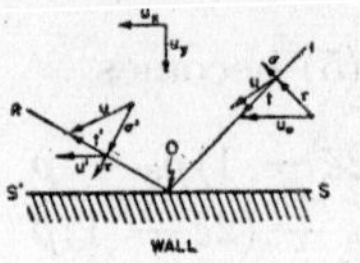

FIG. 2b

It is to be noted that

$$(5') \qquad \eta = \sigma/\tau$$

and that

$$c_0/c = (\xi/\eta)^{1/2}.$$

3. Simple theory of regular reflection in ideal gases. Two plane shock waves which do not have their normals parallel are always in a state of collision. For similar shocks this problem of two-dimensional interaction is equivalent to the reflection of a single shock incident obliquely upon an infinite, plane, rigid wall. The question now is this: Under what conditions does "regular reflection" [13] take place, that is, reflection such that the line of contact between the incident shock and the reflected one moves along the wall? (Inasmuch as the oblique collision of two such shock waves results in an intersection of four shocks, its comparison with the intersection [5, 9] of three shocks is also of interest.) A simple theory of such reflection can be based on the assumption of uniform pressure between the shocks, as well as between each shock and the wall.

Therefore, consider a plane step shock wave I which is moving with constant velocity in an ideal gas of negligible viscosity and thermal conductivity and which is incident at an angle ω upon an infinite, plane, rigid wall $\overline{SS'}$ (Fig. 2a). (It is convenient to consider the gas flow relative to an observer moving with the line of contact 0 (Fig. 2b).) The normal material speed τ in front of the shock I is expressed in terms of ξ in the second equation of (3′), whereas the tangential material speed t along the shock I is determined by

$$(6) \qquad t = \tau \cot \omega.$$

The normal material speed σ behind the shock I, on the other hand, is obtained from the third expression of (3′), which also depends on ξ. Thus the material speed behind the shock I is fixed, too. Resolving this velocity into components along the axes x and y, we have

$$(7) \quad u_x(\xi, \omega) = t \cos \omega + \sigma \sin \omega, \qquad u_y(\xi, \omega) = t \sin \omega - \sigma \cos \omega.$$

Suppose that a plane step shock R is reflected from the wall at an angle ω' (Fig. 2a). It may be characterized by ξ', the ratio of the pressure behind shock R to that in front of it. Now the normal material speed behind shock R is given in terms of ξ' by

$$\tau' = \frac{(\gamma - 1)\xi' + (\gamma + 1)}{(2\gamma[(\gamma + 1)\xi' + (\gamma - 1)])^{1/2}},$$

where the unit of speed is still the speed of sound c in front of that shock. The tangential material speed t' along the shock R is determined by

$$(6') \qquad t' = \tau' \cot \omega'.$$

The normal speed σ' in front of shock R is obtained from the following expression in ξ':

$$\sigma' = \left(\frac{(\gamma + 1)\xi' + (\gamma - 1)}{2\gamma}\right)^{1/2}.$$

Thus the x, y components of the material speed in front of shock R are, respectively, as follows:

$$(7') \quad u_x'(\xi', \omega') = t' \cos \omega' + \sigma' \sin \omega', \qquad u_y'(\xi', \omega') = -t' \sin \omega' + \sigma' \cos \omega'.$$

(It is to be noted that u_x and u_x' are the same function, whereas u_y is merely the negative of u_y'.)

Regular reflection is completely determined by the conditions

$$(8) \qquad u_x(\xi, \omega) = u_x'(\xi', \omega') \quad \text{and} \quad u_y(\xi, \omega) = u_y'(\xi', \omega').$$

Eliminating ω' from these two equations, we obtain the following quadratic expression for σ'^2 in terms of σ and ω:

$$(9) \quad \begin{aligned} &(\gamma + 1)\sigma^4\sigma'^4 - [2(\gamma + 1)^2\sigma^4 + 4\gamma\sigma^2(\sigma^2 - 1)\cos^2 \omega \\ &+ \sigma^2((\gamma - 1)\sigma^2 + 2)^2 \cot^2 \omega]\sigma'^2 + [(\gamma + 1)^2\sigma^4 - 4(\sigma^2 - 1)^2 \cos^2 \omega \\ &+ ((\gamma - 1)\sigma^2 + 2)^2 \cot^2 \omega] = 0. \end{aligned}$$

On the other hand, the condition $u_y = u_y'$ then gives ω' in terms of σ, ω, and σ':

$$\cos \omega' = -\frac{\sigma^{-1} - \sigma}{\sigma'^{-1} - \sigma'} \cos \omega. \tag{10}$$

Some numerical values[3] for $\gamma = 1.40$ are shown here graphically: $\omega'(\omega)$ for given ξ in Fig. 3 and $\xi'(\omega)$ for given ξ in Fig. 4. In general, two solutions ξ', ω'

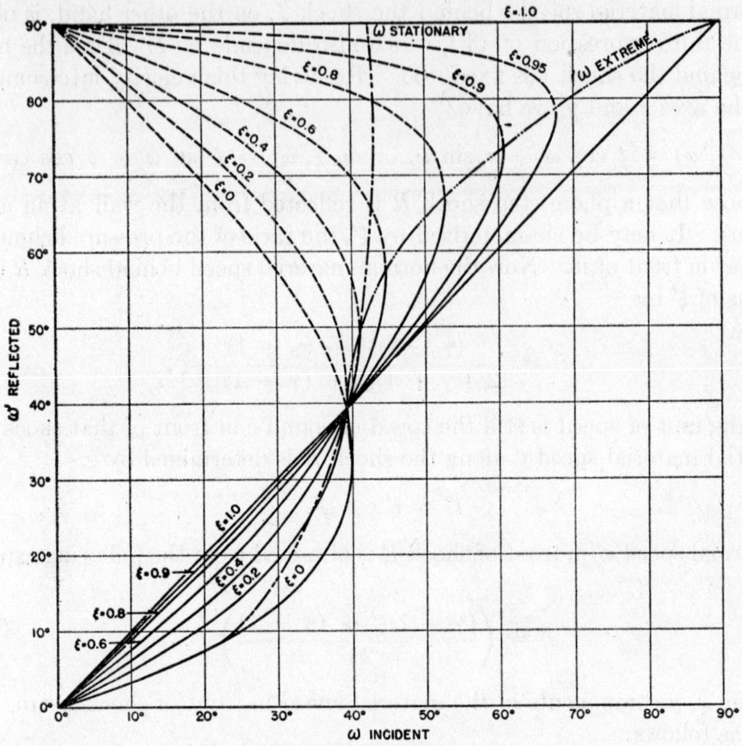

FIG. 3. Ideal gas—$\gamma = 1.40$

are obtained for given ξ, ω; one with high ξ' and high ω', the other with low ξ', and low ω'. There are certain "extreme" values (ω_e), however, for which these two solutions are identical, and beyond which no solutions at all exist. The value of ω_e may be found from the following cubic in $\sin^2\omega_e$ for given $\sigma(\xi)$:

$$\begin{aligned} &16\gamma^2(\sigma^2 - 1)^4 \sin^6 \omega_e - [16(\gamma + 1)^2(\gamma\sigma^2 + 1)(\sigma^2 - 1)^2 + 16\gamma^2(\sigma^2 - 1)^4 \\ &- 8\gamma(\sigma^2 - 1)^2((\gamma - 1)\sigma^2 + 2)^2] \sin^4 \omega_e - [4(\gamma + 1)^2(\sigma^2 - 1)((\gamma - 1)\sigma^2 + 2)^2 \\ &- ((\gamma - 1)\sigma^2 + 2)^4 + 8\gamma(\sigma^2 - 1)^2((\gamma - 1)\sigma^2 + 2)^2] \sin^2 \omega_e \\ &- ((\gamma - 1)\sigma^2 + 2)^4 = 0. \end{aligned} \tag{11}$$

[3] The various numerical results cited in this report are taken from complete tables that are too lengthy to be published. Tables for $\gamma = 1.10$, $\gamma = 1.40$ and $\gamma = 1.67$ were prepared for the Navy Bureau of Ordnance mostly under the Mathematical Tables Project of the Applied Mathematics Panel, NDRC, and are available at the Bureau of Ordnance.

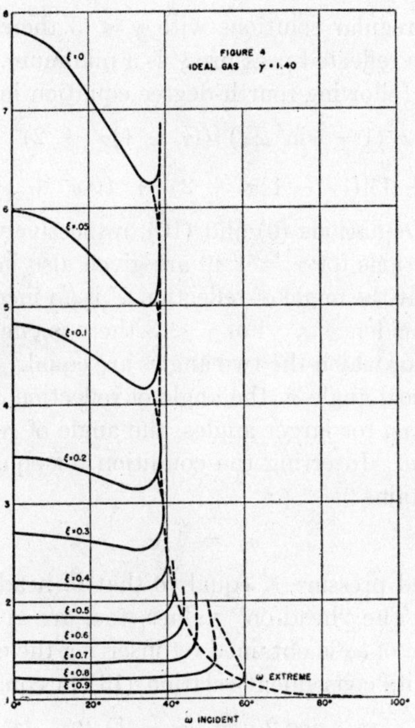

FIG. 4. Ideal gas—$\gamma = 1.40$

TABLE 1

	"Minimum" Solutions			"Head-on" Solutions	"Extreme" Solutions		
	$\gamma = 1.40$			$\gamma = 1.40$	$\gamma = 1.40$		
ξ	ξ'_m	ω_m	ω'_m	ξ'_n	ξ'_e	ω_e	ω'_e
0	6.379	37°.29	22°.04	8.000	6.983	39°.970	32°.97
0.05	5.095	36°.17	22°.64	6.115	5.654	39°.517	35°.29
0.1	4.256	35°.24	23°.17	4.938	4.780	39°.288	37°.47
0.1424	3.739	34°.57	23°.56	4.237	4.237	39°.231	39°.23
0.2	3.210	33°.77	24°.05	3.545	3.677	39°.326	41°.55
0.3	2.574	32°.63	24°.81	2.750	2.991	39°.893	45°.44
0.4	2.141	31°.71	25°.46	2.235	2.509	40°.945	49°.26
0.5	1.825	30°.94	26°.05	1.875	2.145	42°.534	53°.18
0.6	1.584	30°.29	26°.59	1.609	1.852	44°.781	57°.28
0.7	1.392	29°.72	27°.08	1.404	1.607	47°.973	61°.75
0.8	1.237	29°.21	27°.53	1.241	1.392	52°.736	66°.90
0.9	1.109	28°.75	27°.95	1.109	1.194	60°.822	73°.54
0.95	1.052	28°.54	28°.15	1.052	1.098	68°.025	78°.28
1.0	1.000	28°.34	28°.34	1.000	1.000	90°.000	90°.00

Then ξ' and ω' can be obtained as before from equations (9) and (10), respectively. The "extreme" solutions for $\gamma = 1.40$ are given in Table 1.

In the case of the regular solutions with $\gamma < 3$ there is always a particular angle ω_m for which the reflected pressure ξ' is a minimum. The value of ω_m may be obtained from the following fourth-degree equation in $\sin^2 \omega_m$ for given $\sigma(\xi)$:

$$(12) \quad \begin{aligned} 4[(\gamma - 1)\sigma^2 + 2]^2(1 - \sin^2 \omega_m)^2[((\gamma - 1)\sigma^2 + 2)^2 + 4\gamma(\sigma^2 - 1)^2 \sin^4 \omega_m] \\ - (\gamma + 1)[((\gamma - 1)\sigma^2 + 2)^2 + 4(\gamma\sigma^2 + 1)(\sigma^2 - 1) \sin^4 \omega_m]^2 = 0. \end{aligned}$$

Use is again made of equations (9) and (10), respectively, to obtain ξ'_m and ω'_m. The "minimum" solutions for $\gamma = 1.40$ are given also in Table 1.

It is to be noted that the angle of reflection ω' is an increasing monotonic function of the angle of incidence ω. For $\gamma < 3$ there is one particular value of the angle of incidence ω_h for which the two angles are equal. For angles of incidence smaller than this critical angle ω_h the angle of reflection ω' is less than the angle of incidence ω; whereas, for larger angles, the angle of reflection is greater than the angle of incidence. Inserting the condition for equality of these angles in equation (10) one obtains

$$(13) \qquad \sigma'_h = 1/\sigma,$$

which gives a reflected pressure ξ'_h equal to that of head-on reflection for given incident strength ξ. The "head-on" values, too, are given in Table 1 for $\gamma =$ 1.40. The magnitude of ω_h is obtained by inserting the condition for equality of angles together with its consequent relation (13) in equation (9). Thus

$$(14) \qquad \cos 2\omega_h = (\gamma - 1)/2.$$

The value of ω_h is 39°.23 for $\gamma = 1.40$.

For angles of incidence greater than ω_h, the reflected pressure actually exceeds the head-on value. The strongest shock wave for which such regions of reflection exist is given, in general, by

$$(14') \qquad \xi_h = \frac{2\gamma(3 - \gamma)^{1/2} - (3\gamma - 1)}{\gamma + 1}.$$

Thus, for $\gamma = 1.40$, $\xi_h = 0.1424$. This value of ξ, being of special interest, is listed as one of the entries in Tables 1 and 2. The u_y (u_x) curve for $\omega = \omega_h$ is a hyperbola symmetric with respect to the u_x-axis, which makes contact with the envelope of all the hyperbolas only for $\gamma < 1.59307$ [12].

For a given strength ξ_s of the incident shock I there is always one angle of incidence ω_s for which the gas flow through I and R is identical with that in a corresponding solution obtained from the simple theory of three-shock intersections (see below). It is to be noted that for any $\gamma < 3.5931$ this "quasi-stationary" flow corresponds to the regular reflection solution with the low-valued ξ' for weak incident shock ξ, but with the high-valued ξ' for strong incident shocks. The critical strength ξ_{es} that separates these two classes of quasi-stationary solutions is that particular one which is identical with an "extreme" solution. The values of this "extreme-stationary" solution for $\gamma = 1.40$ are:

ξ_{es}	ξ'_{es}	ω_{es}	ω'_{es}
0.4332	2.3783	41°.42	50°.57

Comparisons of the theory of regular reflections of shocks in ideal gases with experiment are given in a number of reports [22, 12, 15]. H. Weyl [26] has shown that the reflection of a shock wave in a fluid satisfying Tammann's equation of state has characteristics similar to those in an ideal gas. The authors [17] themselves have considered waterlike substances in detail.

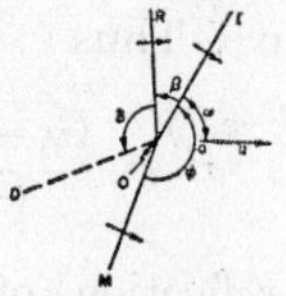

FIG. 5a

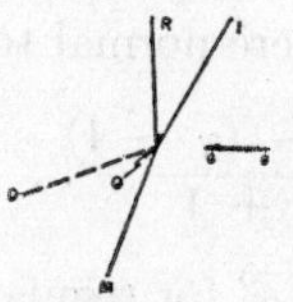

FIG. 5b

4. **Simple theory of three-shock intersections in ideal gases.** Three plane shocks I, R, M intersect along a line O (normal to the plane of the paper in Fig. 5a), which moves into still gas in the direction of $\overrightarrow{aa}$. The simple theory [16, 5, 9] requires that each region between adjacent shocks have uniform pressure. (It can be applied also as a first approximation in the neighborhood of an intersection where the pressure in each region is not quite uniform.) Consideration of the gas flow relative to the moving intersection-line O indicates that one of these regions must have a plane along which there is a discontinuity in tangential material speed (hence in density, too). Thus in Fig. 5b the three shocks are regarded as stationary and the gas as moving initially along aa and then parallel to the discontinuity surface D (a "slipstream"). We shall use this frame of reference henceforth.

The specification of two parameters in addition to the physical state of an ideal gas of negligible viscosity and thermal conductivity in one of the uniform regions is sufficient to determine the solution. For computational convenience we choose as parameters ξ, the ratio of the pressure ahead of shock I to that behind it, and ω, the angle between the shock front I and the direction of motion

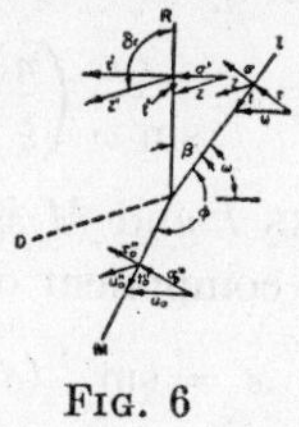

FIG. 6

of the line of intersection O. We fix the remaining quantities defining the state in front of shock I, but these need not enter directly in our computation. From ξ we obtain immediately the other characteristics of the shock I, such as η, σ, τ (cf. equations (3′), (5′)).

Now t, the tangential material speed along the shock front I, is unchanged across the shock and is given by (cf. Fig. 6) $t = \tau \cot \omega$. Thus the velocity of the gas flow in the region between the shocks I and R is determined. Suppose β, the angle between these shocks I and R, is known. Resolving the material velocity into components σ' normal to the shock R and t' tangential to it, we obtain

$$\sigma' = t \sin \beta + \sigma \cos \beta \quad \text{and} \quad t' = t \cos \beta - \sigma \sin \beta, \tag{15}$$

where a positive value of t' signifies the direction along R toward O. Now σ' determines ξ', the ratio of the pressure behind shock R to that ahead of it and also τ', the gas velocity there normal to it. Thus

$$\xi' = \frac{2\gamma\sigma'^2 - (\gamma - 1)}{\gamma + 1}, \qquad \tau' = \frac{(\gamma - 1)\sigma'^2 + 2}{(\gamma + 1)\sigma'}.$$

Hence the direction δ_R ($= \omega'$ for regular reflection) of gas flow behind the shock R is given by

$$\delta_R = \tan^{-1}(\tau'/-t'). \tag{16}$$

For the purpose of considering the gas flow across the remaining shock M it is convenient to normalize quantities with respect to the region ahead of M instead of that behind I, that is, p_0, ρ_0, c_0. Let ξ_0'' represent the ratio of the pressure behind shock M to that ahead of it. Thus

$$\xi_0'' = p''/p_0 = \xi'/\xi. \tag{17}$$

This value of ξ_0'' fixes the normal gas material speed σ_0'' in front of the shock M, that is,

$$\sigma_0'' = \left(\frac{(\gamma + 1)\xi_0'' + (\gamma - 1)}{2\gamma}\right)^{1/2},$$

as well as the normal material speed τ_0'' behind it, that is,

$$\tau_0'' = \frac{(\gamma - 1)\xi_0'' + (\gamma + 1)}{(2\gamma[(\gamma + 1)\xi_0'' + (\gamma - 1)])^{1/2}} = \frac{(\gamma - 1)\sigma_0''^2 + 2}{(\gamma + 1)\sigma_0''}.$$

But the material speed u_0 in the region in front of shock M is given in terms of the sound speed c_0 for that region by the equation

$$u_0 = \frac{\tau}{\sin \omega}\left(\frac{\eta}{\xi}\right)^{1/2}. \tag{18}$$

Thus the angle ϕ between shocks I and M is determined by the fact that the normal material velocity σ_0'' is a component of u_0, that is,

$$\phi - \omega = \sin^{-1}(\sigma_0''/u_0). \tag{19}$$

Hence the tangential material speed t_0'' along the shock M is given by

$$t_0'' = \sigma_0'' \cot(\phi - \omega), \tag{20}$$

where a positive value of t_0'' signifies the direction along M toward O, so that the direction of gas-flow behind the shock M is determined by

$$\delta_M = 2\pi - [\phi + \beta + \tan^{-1}(\tau_0''/-t_0'')]. \tag{21}$$

Thus if the value of β has been chosen correctly, a solution will exist for $\delta_M = \delta_R$.

As in the case of regular reflection, there is usually more than one set of real solutions for a given set of parameters. That this is the case may be deduced from considerations of limiting conditions, which indicate that there are, at least, two families of solutions. Thus in the case of weak reflected shocks, that is, $\xi' \to 1$, after much algebraic manipulation one obtains the following quadratic equation for the limiting angle β_1'

$$\begin{aligned}(22)\quad & 2[(2\gamma - 1)\sigma^4 + 2\sigma^2 + 1]\cos^2\beta_1' - 2\sigma[(\gamma - 1)\sigma^4 + 4\gamma\sigma^2 - (\gamma - 5)]\cos\beta_1' \\ & + 3(\gamma - 1)\sigma^4 + 2(\gamma + 3)\sigma^2 - (\gamma - 1) = 0.\end{aligned}$$

It is to be noted that ϕ_1' approaches π as a limit, whereas the limiting angles ω_1', δ_1' are given by

$$\tan\omega_1' = \frac{\tau}{\csc\beta_1' - \sigma\cot\beta_1'}, \qquad \text{and} \qquad \cot\delta_1' = \sigma\csc\beta_1' - \cot\beta_1'.$$

On the other hand, for strong reflected shocks, that is, $\xi' \to \infty$, the limiting angle β_∞' is found from a quadratic equation in $\sin^2\beta_\infty'$:

$$(23)\qquad 4\gamma\sin^4\beta_\infty' - (\gamma + 1)^2(\eta + 1)\sin^2\beta_\infty' + (\gamma + 1)^2\eta = 0.$$

In this case the two physically possible values of β_∞' are supplementary and ω_∞' approaches zero as a limit. Furthermore

$$\tan\phi_\infty' = -(\gamma + 1)/(\gamma - 1)\cot\beta_\infty', \qquad \text{and} \qquad \cot\delta_\infty' = \tan\phi_\infty'.$$

Finally, for the limiting value $\xi \to 1$ one obtains the following cubic equation for the limiting angle β_1 which is supplementary to ϕ_1:

$$\begin{aligned}(24)\quad & 8\sigma'(\gamma\sigma'^2 + 1)(\sigma'^2 - 1)\cos^3\beta_1 - 2[2\gamma\sigma'^6 + (2\gamma^2 + 5\gamma + 5)\sigma'^4 \\ & - 2(2\gamma + 1)\sigma'^2 + (\gamma - 1)]\cos^2\beta_1 + 2\sigma'[(\gamma^2 + 3)\sigma'^4 + 4(\gamma^2 + 2\gamma - 1)\sigma'^2 \\ & - (\gamma^2 - 5)]\cos\beta_1 + 2(\gamma - 1)\sigma'^6 - (3\gamma + 5)(\gamma - 1)\sigma'^4 \\ & - 2(\gamma^2 + 5\gamma + 2)\sigma'^2 + (\gamma + 3)(\gamma - 1) = 0.\end{aligned}$$

Also,

$$\cot\omega_1 = \sigma'\csc\beta_1 - \cot\beta_1 \qquad \text{and} \qquad \cot\delta_1 = \frac{\csc\beta_1 - \sigma'\cot\beta_1}{\tau'}.$$

As γ approaches unity, the limiting configurations become quite simple. For hypothetical values of γ greater than 1.67, which are of use in the discussion of waterlike substances [13, 17], some of the limiting conditions just found do not exist. For example, in the case of one of the families of solutions there is no sonic reflected shock ($\xi_1' \to 1$) for very weak initial shocks ($\xi \to 1$) if γ is greater than 1.67, whereas there is no sonic reflected shock for very strong initial shocks ($\xi \to 0$) if γ is greater than 2.

Numerical results for $\omega(\beta)$ are shown graphically in Fig. 7 for $\gamma = 1.40$ and for given values of ξ, namely, $\xi \to 1$, $\xi = 0.9, 0.7, 0.5, 0.3, 0.1, 0.0$. It is to be noted that there are two families of solutions; also, that there are certain values of the angles for which no three-shock configurations are possible, as well as other values for which the two families of solutions overlap. The two families can best be distinguished by the relative magnitudes of the angle δ. For $\gamma = 1.40$ the values for δ lie entirely above 135° for one family of solutions and below 135° for the second family.

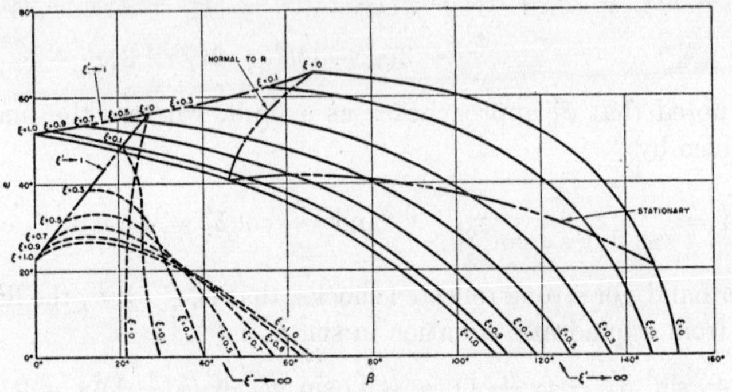

FIG. 7. Ideal gas—$\gamma = 1.40$
Three-shock configurations

Of particular interest are the cases where the gas flows across one of the shocks normally. This condition cannot be realized for shock I. In this instance the tangential component of the material velocity along I is zero so that the normal component σ' across shock R (and consequently ξ' the strength of R) would be less than unity, inasmuch as $\sigma < 1$ and $\cos\beta \leqq 1$ (cf. the first equation of (15)).

Let us consider the case where the material flow across shock M is normal to it. Now $\phi - \omega = \pi/2$ in this instance, so that

$$\sigma_0'' = u_0 .$$

This relation gives ξ', and hence σ', as a function of ξ and ω. Thus the value of β is determined by the first equation of (15) and δ_R, in turn, by equation (16). But in this case $\delta_R = \pi - (\omega + \beta)$ so that ω, and consequently β, must be selected to obtain consistent values of δ_R for given ξ. The solutions thus obtained are the "quasi-stationary" ones related in each instance to a case of regular reflection. It is possible to obtain $\sigma_s'^2$ algebraically from the following quadratic equation

$$(25)\quad 2\sigma^2\sigma_s'^4 - (2\sigma^4 + (\gamma - 1)\sigma^2 + 2)\sigma_s'^2 - (\sigma^2 - 1)((\gamma - 1)\sigma^2 + 2) = 0,$$

and then ω_s from

$$\tan^2 \omega_s = \frac{(\gamma - 1)\sigma^2 + 2}{(\gamma + 1)(\sigma_s'^2 - 1)} .$$

The solutions are given in Table 2 for $\gamma = 1.40$; they are indicated by the word "stationary" in Fig. 7.

TABLE 2

"Stationary" Solutions

	$\gamma = 1.40$		
ξ	ξ'_s	ω_s	ω'_s
0	7.270	21°77	8°37
0.05	5.466	25°42	11°60
0.1	4.381	28°74	15°46
0.1424	3.769	31°30	19°21
0.2	3.211	34°36	24°96
0.3	2.674	38°38	36°02
0.4	2.425	40°87	47°09
0.5	2.317	42°21	57°13
0.6	2.280	42°83	65°83
0.7	2.283	43°01	73°29
0.8	2.309	42°94	79°69
0.9	2.349	42°71	85°20
0.95	2.374	42°56	87°68
1.0	2.400	42°39	90°00

The case where the material flows normally across the shock R can be simply related to the quasi-stationary state [9]. In this instance $\delta = \pi/2$ so that $t' = 0$. Hence by the second equation of (15) β is also fixed; σ' is then given by the first equation of (15), σ_0'' by (17) and the Rankine-Hugoniot relations [11], u_0 by (18), and finally ϕ is determined by (19). These solutions are indicated by the phrase "normal to R" in Fig. 7. It is of interest that both the quasi-stationary configurations and those for flow normal to R occur only in one family of solutions.

Extensive tables and charts ($\gamma = 1.10, 1.40, 1.67$) for the various quantities occurring in the simple theory of regular reflection and of three-shock configurations are contained in the Navy Bureau of Ordnance Report ERR 13 [16].

5. Modification of the simple theory of shock intersections. The so-called simple theory of three-shock intersections in an ideal gas has been found inadequate to explain a large class of three-shock configurations obtained from experiment [22, 12, 15, 10]. Also, complete agreement [10] does not exist between observed and theoretically computed angles in the case of the regular reflection of strong shocks. Now this theory is based on the assumption that each region bounded by shock waves has uniform pressure. In order to facilitate the analysis of experimental data, therefore, it is necessary to consider the possibility of a more generalized flow pattern in the neighborhood of an intersection. In this section consideration will be given to the modification of one or more regions between adjacent shock waves by the inclusion of a Prandtl-Meyer wave [24, 15, 7],[4] which is a zone of continuous pressure variation of the type encountered in

[4] R. J. Seeger, OSRD4943.

the bending of supersonic flow around a corner. The pressure, density, and material velocity in the interior of a Prandtl-Meyer (PM) wave is constant along any radius $\overline{OS}$ (Fig. 8), but varies continuously with the angle θ, measured from an initial radius $\overline{OR}$ along which the flow speed is sonic and normal to $\overline{OR}$. If M is the ratio of the material speed to the local speed of sound at any point on

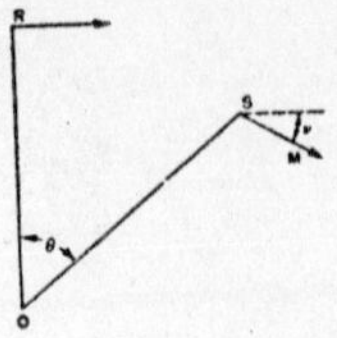

FIG. 8

radius $\overline{OS}$, then the angle θ and the pressure ratio $P = Ps/Pr$ (Ps and Pr are the pressures on $\overline{OS}$ and $\overline{OR}$, respectively) are given by the expressions below

$$\theta = \left(\frac{\gamma + 1}{\gamma - 1}\right)^{1/2} \tan^{-1}\left[\left(\frac{\gamma - 1}{\gamma + 1}\right)^{1/2} (M^2 - 1)\right], \tag{26}$$

$$P = \left[\cos\left(\frac{\gamma - 1}{\gamma + 1}\right)^{1/2} \theta\right]^{2\gamma/(\gamma-1)}.$$

Also, the angular rotation ν of the material-velocity vector in going from a point on $\overline{OR}$ to $\overline{OS}$ is given by

$$\nu = \theta - \sec^{-1} M. \tag{27}$$

From equation (27) it follows that the material speed normal to $\overline{OS}$ is always equal to the local speed of sound, which is one of the characteristic properties of Prandtl-Meyer waves. It follows also that the material speed is everywhere supersonic.

We shall consider first the case of regular reflection of a plane shock wave from a rigid wall. It is evident that no PM flow is possible in the region between the reflected shock and the wall. This is a consequence of the fact that the material velocity behind the reflected shock and normal to it is subsonic; hence it must be subsonic in a direction perpendicular to any line in that region. It follows also from similar considerations that no shock wave can follow a PM rarefaction wave. It is, therefore, possible to construct a flow pattern only with a PM compression wave in the region between the incident and reflected shocks. The effect produced by the introduction of such a PM compression wave is to decrease the magnitude of the extreme incident angle ω_e.

The mathematical relations that must be satisfied by the various physical quantities are given by the system of equations (28). M_1 and M_2 are the material speeds prior to and following the PM wave, respectively; while ΔM, $\Delta\nu$, ΔP are the total changes experienced by M, ν, P while passing through this wave. The quantities τ, σ, and τ', σ' represent the material speeds normal to the incident and reflected shocks, respectively; t and t' are the speeds parallel to the shock fronts. All speeds pertaining to the incident shock are normalized with respect to the speed of sound in the region immediately behind the incident shock, while all speeds pertaining to the reflected shock are normalized with respect to the speed of sound in the region directly in front of the reflected shock. The rela-

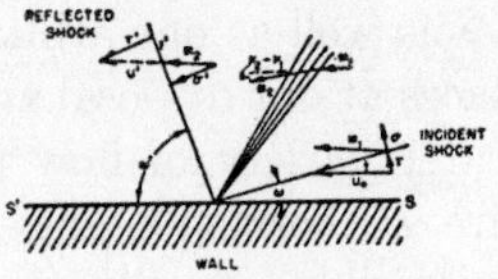

Fig. 9

tions for σ, τ, η are the Rankine-Hugoniot equations for shock waves, and hold equally well for σ', τ', η'. It is now necessary to prescribe three parameters in order to obtain a solution. For instance, we may prescribe ω, ξ, M_2.

$$\sigma = \left(\frac{(\gamma + 1)\xi + (\gamma - 1)}{2\gamma}\right)^{1/2}; \quad \tau = \frac{(\gamma - 1)\sigma^2 + 2}{(\gamma + 1)\sigma};$$

$$\eta = \frac{\sigma}{\tau} = \frac{(\gamma + 1)\xi + (\gamma - 1)}{(\gamma - 1)\xi + (\gamma + 1)},$$

$$(28) \qquad t = \tau \cot \omega, \qquad M_1 = (\sigma^2 + t^2)^{1/2},$$

$$t' = M_2 \cos \lambda; \qquad \sigma' = M_2 \sin \lambda;$$

$$\lambda = \omega' + \omega + |\nu_2 - \nu_1| - \tan^{-1}(\sigma/t),$$

$$\omega' = \tan^{-1} \tau'/t'.$$

The velocity vectors corresponding to the speeds τ, σ, t, u_1 and τ', σ', t', u' are indicated schematically in Fig. 9.

Considerations similar to those of regular reflection [22] limit the existence of PM waves for the case of three-shock configurations to the following possibilities:

(a) A compression PM wave between the incident and reflected shocks,

(b) A compression, rarefaction, or compression-rarefaction PM wave between the reflected shock and the density discontinuity,

(c) A compression PM wave in front of the Mach shock.

A system of equations similar to those for three-shock configurations given in §3 must be satisfied in each case, in addition to equations (26) and (27) which hold in the interior of the PM wave and to the Rankine-Hugoniot equations which relate the physical states on both sides of a shock wave. For instance, for a PM wave in the region between the reflected shock and density discontinuity the following modifications of the simple three-shock-configuration equations must be made. Equation (29) gives the material speed (Mach number) M_1 behind the reflected shock

$$M_1 = (\eta'/\xi')^{1/2}(\tau'^2 + t'^2)^{1/2}. \tag{29}$$

Equation (17) is replaced by

$$\xi'' = (P_2/P_1)(\xi'/\xi). \tag{17'}$$

Since the physical conditions behind a PM wave are not determined unless the conditions in front of the wave as well as one arbitrary parameter are specified, the inclusion of a PM wave places at our disposal an additional parameter which may be used to obtain a greater variety of flow patterns. As in the case of regular reflection, the pressure P_2 corresponding to a material speed (Mach number) M_2 may be chosen arbitrarily. Finally, the equation $\delta_M = \delta_R$ is replaced by the analogous equation $\delta_M = \delta_R + |\nu_2 - \nu_1|$.

In addition to the limitations discussed above there are a number of physical restrictions to the solution of three-shock Prandtl-Meyer configurations which apply particularly to the region between the reflected shock and density discontinuity. First, the material velocity must be supersonic in a region where a PM wave can be fitted. Secondly, the direction of flow must be in the direction of the intersection line, rather than away from it.

6. **Refraction of a shock wave at a free surface.** A particularly interesting wave pattern is that which may be expected to result from the interaction of a shock wave with a free surface. Consider two media of different physical properties separated by an interface. If a shock wave originating in one of these media impinges upon the second medium at an arbitrary angle ω, then in general it may be expected that a shock wave will be transmitted into the second medium, while a second wave will be reflected either as a shock or as a rarefaction wave. In the first case the resulting pattern will be a three-shock configuration of a somewhat more general type than that discussed in §3. In the second case a new type of configuration will result, consisting of two shock waves and a Prandtl-Meyer rarefaction wave. We shall limit our present discussion to ideal gases. Let γ_0, c_0, p_0, and γ_1, c_1, p_1 be the ratios of the specific heats, the sound speeds, and the pressures, respectively, pertaining to the two regions (it will be noted that $p_0 = p_1$). Let a plane step shock wave of strength ξ, traveling in the region (γ_0, c_0, p_0), strike the interface at an angle ω. Then in the case of a reflected shock wave there will result a three-shock configuration defined by a set of equations analogous to those of §3, with the following modifications. Equation (18) will be replaced by

$$u_0 = \frac{c_0}{c_1} \frac{\tau}{\sin \omega} \left(\frac{\eta}{\xi}\right)^{1/2}, \tag{18'}$$

and the Rankine-Hugoniot relations involving the transmitted shock will have γ replaced by γ_1; while those of the incident and reflected shocks will have γ replaced by γ_0. On the other hand, to obtain the second type of configuration we must make the following changes, in addition to the above. Equations (15),

(16), and (17) pertaining to the reflected shock wave are replaced by equivalent relations for a Prandtl-Meyer wave,

$$M_1 = (\sigma^2 + t^2)^{1/2},$$

$$\theta_1 = \left(\frac{\gamma_0 + 1}{\gamma_0 - 1}\right)^{1/2} \tan^{-1}\left[\left(\frac{\gamma_0 - 1}{\gamma_0 + 1}\right)^{1/2} (M_1^2 - 1)^{1/2}\right];$$

$$\theta_2 = \left(\frac{\gamma_0 + 1}{\gamma_0 - 1}\right)^{1/2} \tan^{-1}\left[\left(\frac{\gamma_0 - 1}{\gamma_0 + 1}\right)^{1/2} (M_2^2 - 1)^{1/2}\right],$$

(15′) $$P_1 = \left[\cos\left(\frac{\gamma_0 - 1}{\gamma_0 + 1}\right)^{1/2} \theta_1\right]^{2\gamma_0/(\gamma_0 - 1)}; \quad P_2 = \left[\cos\left(\frac{\gamma_0 - 1}{\gamma_0 + 1}\right)^{1/2} \theta_2\right]^{2\gamma_0/(\gamma_0 - 1)},$$

$$\nu_1 = \theta_1 - \sec^{-1} M_1; \qquad \nu_2 = \theta_2 - \sec^{-1} M_2,$$

(16′) $$\lambda_R = \pi - \tan^{-1}(\sigma/t) + |\nu_2 - \nu_1|,$$

(17′) $$\xi'' = P_2/P_1 \xi.$$

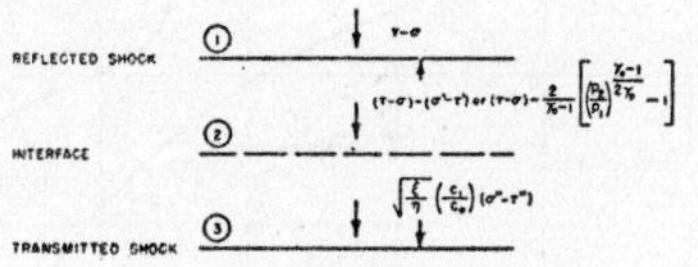

FIG. 10

Equation (21) is replaced by

(21′) $$\lambda_M = 2\pi - \phi - \tan^{-1}(\tau''/-t_0'').$$

Finally, the relation $\lambda_R = \lambda_M$ is substituted for the equation $\delta_R = \delta_M$.

The above systems of equations for the two types of refraction patterns constitute a complete solution of the problem on the basis of our simplified assumptions. They contain all possible configurations resulting from the collision of a shock wave with a free surface. An exhaustive analysis, however, based on these equations is rather difficult, due to their algebraic complexity. For instance, for a fixed set of values of the parameters it is possible to obtain a number of solutions. Also it is not possible to tell a priori what type of refraction pattern to expect for given conditions. In order to obtain an insight into the character of these solutions it is desirable, therefore, to investigate first certain special cases whose solutions are predictable from other considerations. Two such cases are the refraction of a shock at normal incidence (that is, $\omega = 0$) and the refraction of an acoustic wave (that is, $\xi = 1$) at any angle of incidence.

For the case of normal incidence we can derive the solution to the above problem directly by a rather simple argument. To fix our ideas, assume that the flow on both sides of the interface was originally at rest. After the interaction has taken place, the material velocity on both sides of the density discontinuity must remain equal. The material speeds in regions (1), (2), and (3) (Fig. 10), normalized with respect to the speed of sound in region (1), are $(\tau - \sigma)$,

$(\tau - \sigma) - (\sigma' - \tau')$, $(\xi/\eta)^{1/2}(c_1/c_0)(\sigma'' - \tau'')$, respectively, for a reflected shock wave; while for a reflected (simple) rarefaction wave these speeds are $(\tau - \sigma)$, $(\tau - \sigma) - (2/(\gamma_0 - 1))[(P_2/P_1)^{(\gamma_0-1)/2\gamma_0} - 1]$ and $(\xi/\eta)^{1/2}(c_1/c_0)(\sigma'' - \tau'')$, respectively. Equating the speeds in regions (2) and (3) we obtain

$$(30) \qquad (\tau' - \sigma') + (\tau - \sigma) + (\xi/\eta)^{1/2}(c_1/c_0)(\tau'' - \sigma'') = 0$$

for a reflected shock wave; and

$$(30') \qquad \frac{2}{\gamma_0 - 1}\left[1 - \left(\frac{P_2}{P_1}\right)^{(\gamma_0-1)/2\gamma_0}\right] + (\tau - \sigma) + (\xi/\eta)^{1/2}(c_1/c_0)(\sigma'' - \tau'') = 0$$

for a reflected rarefaction wave.

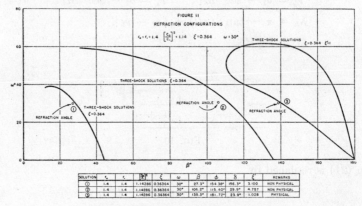

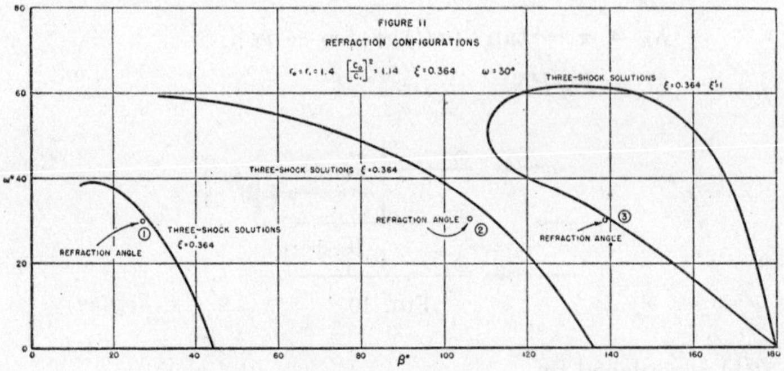

SOLUTION	r_0	r_1	$\left[\frac{c_0}{c_1}\right]^2$	ξ	ω	β	ϕ	δ	ξ'	REMARKS
①	1.4	1.4	1.14286	0.36364	30°	27.3°	154.38°	156.3°	3.100	NON PHYSICAL
②	1.4	1.4	1.14286	0.36364	30°	106.2°	115.40°	29.9°	4.757	NON PHYSICAL
③	1.4	1.4	1.14286	0.36364	30°	139.3°	181.72°	23.9°	1.028	PHYSICAL

FIG. 11. Refraction configurations

It can be shown that the general equations for the two types of refraction patterns also approach the above limits for $\omega \to 0$. For the case of a reflected shock wave this limit is approached in that branch for which also $\sin \beta \to 0$. For both types of patterns we may expect that the physically plausible solution is that one which is connected continuously to the solution given by equations (30) and (31).

It is interesting to note that neither of the two branches found to satisfy the equations for the three-shock configuration problem in §3 (equivalent to the refraction problem for the special case $\gamma_0 = \gamma_1$, $c_0/c_1 = 1$) is a continuation of the solution to equation (30). As a matter of fact, the value of ξ' (the strength of the reflected shock) tends to infinity for both these solutions. On the other hand, a third branch exists for the solution of the refraction problem, which is apparently the physically likely solution and which approaches the solution to equation (30) for $\omega = 0$. This branch degenerates to the trivial solution $\xi' = 1$ for three-shock configurations ($\gamma_0 = \gamma_1$, $c_0/c_1 = 1$). This property is illustrated in Fig. 11, where the three $\omega(\beta)$ solutions of the refraction problem for $\gamma_0 = \gamma_1 = 1.40$, $(c_0/c_1)^2 = 1.14$, $\xi = 0.364$, $\omega = 30°$ are plotted against a background of the corresponding families of solutions for the three-shock configuration problem for the same parameters (but with $(c_0/c_1) = 1$).

In the case of an incident acoustic wave ($\xi = 1$) Snell's law, namely,

$$\frac{\sin \omega}{\sin (\phi - \omega)} = \frac{c_0}{c_1} \tag{31}$$

is known to be physically satisfactory. We can verify by direct substitution in the equations for the general refraction patterns that this equation is satisfied for $\xi = 1$. Here again we can identify the physically plausible branch in the neighborhood of $\xi = 1$ by finding that branch which is continuously connected with the solution given by equation (31).

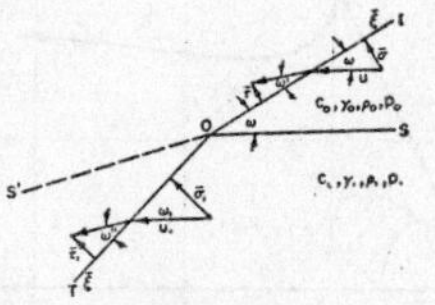

FIG. 12

Still another special type of refraction pattern of particular interest is that for which a reflected wave is entirely absent. This is the transition point between a refraction configuration with a reflected shock and one with a reflected rarefaction. It is analogous to the case of equal acoustic impedances ($\rho_0 c_0 = \rho_1 c_1$) on both sides of an interface for normal incidence of an acoustic wave (or for normal incidence of a shock wave when $\gamma_0 = \gamma_1$). Fortunately, this result may be easily obtained, and the basic equation may be reduced to a quadratic in $S = \sin^2\omega$.

In Figure 12, OI is a shock wave intersecting the interface OS at O. OT is the transmitted shock, and OS' is the direction of flow behind either shock wave. If no reflected wave is produced, it follows from the equality of pressures on both sides of OS' that the incident and transmitted waves are of equal strength. Let ξ be the ratio of pressure across either shock, normalized with respect to the region in front (to the right). The following equations must then be satisfied. The notation is essentially that used in the derivation of the general problem.

$$\bar{\sigma} = \left(\frac{(\gamma_0 + 1)\xi + (\gamma_0 - 1)}{2\gamma_0}\right)^{1/2}; \qquad \bar{\eta} = \frac{(\gamma_0 + 1)\xi + (\gamma_0 - 1)}{(\gamma_0 - 1)\xi + (\gamma_0 + 1)},$$

$$u = \frac{\bar{\sigma}}{\sin \omega}; \qquad u_1 = \frac{\bar{\sigma}}{\sin \omega}\left(\frac{c_0}{c_1}\right),$$

$$\bar{\sigma}_1 = \left(\frac{(\gamma_1 + 1)\xi + (\gamma_1 - 1)}{2\gamma_1}\right)^{1/2}; \qquad \bar{\eta}_1 = \frac{(\gamma_1 + 1)\xi + (\gamma_1 - 1)}{(\gamma_1 - 1)\xi + (\gamma_1 + 1)}, \tag{32}$$

$$\omega_1 = \sin^{-1}(\bar{\sigma}_1/u_1); \qquad \tan \omega' = \tan \omega/\bar{\eta}; \qquad \tan \omega'' = \tan \omega_1/\bar{\eta}_1,$$

$$\omega_1 - \omega'' = \omega - \omega'.$$

By substituting the values for ω_1, ω', and ω'' in the last of equations (32) we obtain after some algebraic manipulations the following equivalent relation,

$$(32')\qquad 4\gamma_1(\bar{\xi}^2 - 1)\left[\frac{\gamma_0\,\alpha_1}{a\gamma_1\alpha_0} - 1\right]S^2 + \left[\alpha_0\,\alpha_1\left(a - \frac{1}{a}\right) + 4(\gamma_1 - \gamma_0)(\bar{\xi}^2 - 1)\right]S + \alpha_0[\alpha_0 - a\alpha_1] = 0$$

where

$$a = \gamma_1/\gamma_0(c_0/c_1)^2; \qquad \alpha_0 = (\gamma_0 + 1)\bar{\xi} + (\gamma_0 - 1);$$
$$\alpha_1 = (\gamma_1 + 1)\bar{\xi} + (\gamma_1 - 1); \qquad S = \sin^2\omega.$$

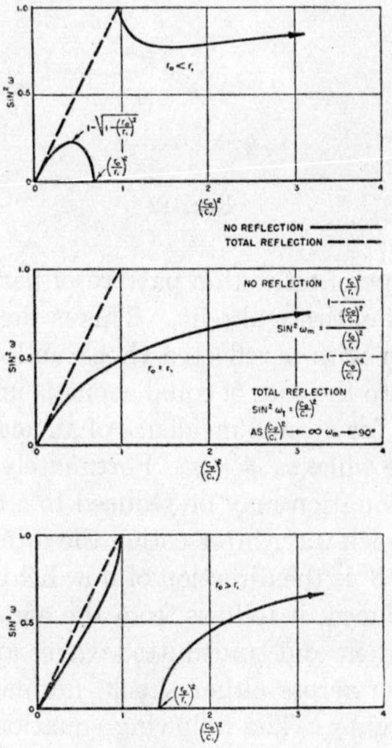

FIG. 13. Refraction of acoustic wave

For normal incidence this equation reduces to

$$(33)\qquad \frac{(\gamma_0[(\gamma_0 + 1)\bar{\xi} + (\gamma_0 - 1)])^{1/2}}{c_0} = \frac{(\gamma_1[(\gamma_1 + 1)\bar{\xi} + (\gamma_1 - 1)])^{1/2}}{c_1}.$$

This equation is equivalent to the relation

$$(33')\qquad [\tau(\xi) - \sigma(\xi)] + (\xi/\eta)^{1/2}(c_1/c_0)[\sigma''(\xi) - \tau''(\xi)],$$

which may be obtained directly from equation (30) or (30′) by setting $\xi' = 1$ or $P_2/P_1 = 1$. This, in turn, is equivalent to the condition that the material velocity in region (1) is equal to that of region (3) (Fig. 10).

For the case of an acoustic incident wave equation (32′) reduces to [19]

$$\sin^2 \omega_t = \frac{1 - (\gamma_0/\gamma_1)^2(c_1/c_0)^2}{1 - (\gamma_0/\gamma_1)^2(c_1/c_0)^4}. \tag{34}$$

For such a wave the angle for total reflection (transmitted wave normal to interface) is given by the relation

$$\sin^2 \omega_r = (c_0/c_1)^2. \tag{31′}$$

In Figure 13 we plot both $\sin^2 \omega_t$ and $\sin^2 \omega_r$ as a function of $(c_0/c_1)^2$, for the three cases $\gamma_0 < \gamma_1$, $\gamma_0 = \gamma_1$, $\gamma_0 > \gamma_1$. It will be noted that in all cases $\omega_r > \omega_t$ for real values of ω. This property is of considerable interest, since total reflection can take place only when the reflected wave is a shock. (In the case of a reflected Prandtl-Meyer rarefaction wave the flow in the upper region would be bent towards the intersection line both by the incident wave and the reflected wave and can thus never be parallel to the interface.) Hence in order that total reflection be feasible the refraction pattern for angles $\omega > \omega_t$ must be of the reflected shock-wave variety.

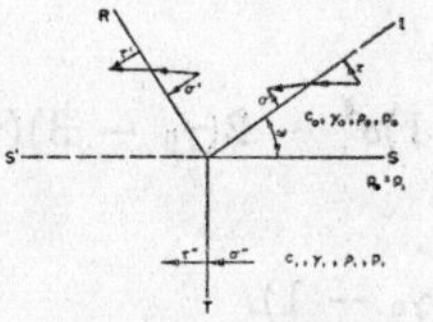

FIG. 14

To derive the equations for total reflection we must consider only the case of a reflected shock wave. We can obtain such a system of equations by setting $(\phi - \omega) = \pi/2$ in the system of equations for the general refraction configuration. It is possible, however, to deduce a cubic equation in σ'^2, analogous to the well-known quadratic equation derived independently by K. Friedrichs and by J. von Neumann (equation (25)) for the case of a stationary shock configuration [13]. To do this we consider the two regions on both sides of the interface separately (Fig. 14). The solution of the shock configuration in the upper half is identical with the case of regular reflection (equation (9)) from a rigid wall, which replaces SS' in the diagram.

The condition that must be satisfied in the lower half is simply the equality of pressure in the region behind the transmitted shock to that behind the reflected shock. Since the flow in front of the transmitted shock has no vertical component the relation

$$\sigma'' = \frac{\tau}{\sin \omega}\left(\frac{\eta}{\xi}\right)^{1/2}\left(\frac{c_0}{c_1}\right) \tag{18″}$$

must hold. This is equivalent to

$$(18''') \qquad \csc^2 \omega = \frac{\gamma_0(\gamma_1 + 1)\sigma'^2 + \gamma_0(\gamma_1 - 1)\sigma^2 - \gamma_1(\gamma_0 - 1)}{\gamma_1(c_0/c_1)^2[(\gamma_0 - 1)\sigma^2 + 2]}.$$

We may eliminate ω in equation (9) by substituting its value from (18'''), thus obtaining a cubic expression in σ'^2, which may be written in the form,

$$(35) \qquad A\sigma'^6 + B\sigma'^4 + C\sigma'^2 + D = 0,$$

where

$$A = \frac{\gamma_0(\gamma_1 + 1)(\gamma_0 + 1)^2\sigma^4}{\gamma_1(c_0/c_1)^2} - \frac{\gamma_0^2(\gamma_1 + 1)^2}{\gamma_1^2(c_0/c_1)^4}\, g,$$

$$B = \frac{(\gamma_0 + 1)^2\sigma^4}{\gamma_1(c_0/c_1)^2}\, h - \frac{\gamma_0(\gamma_1 + 1)}{\gamma_1(c_0/c_1)^2}\, f + \frac{[\gamma_0^2(\gamma_1 + 1)^2 - 2\gamma_0(\gamma_1 + 1)\sigma^2 h]}{\gamma_1^2(c_0/c_1)^4}\, g,$$

$$C = 4\gamma_0\sigma^2(\sigma^2 - 1)^2 g - \frac{fh}{\gamma_1(c_0/c_1)^2} + \frac{[2\gamma_0(\gamma_1 + 1) - \sigma^2 h]}{\gamma_1^2(c_0/c_1)^4}\, gh + \frac{4(\sigma^2 - 1)\gamma_0(\gamma_1 + 1)}{\gamma_1(c_0/c_1)^2}\, g,$$

$$D = g\left[4(\sigma^2 - 1)^2 + \frac{4(\sigma^2 - 1)}{\gamma_1(c_0/c_1)^2}\, h + \frac{1}{\gamma_1^2(c_0/c_1)^4}\, h^2\right]$$

and where

$$f = -\sigma^2[(\gamma_0^2 - 6\gamma_0 + 1)\sigma^4 - 2(\gamma_0 - 3)(\gamma_0 - 1)\sigma^2 - 4(\gamma_0 - 1)],$$

$$g = (\gamma_0 - 1)\sigma^2 + 2,$$

$$h = \gamma_0(\gamma_1 - 1)\sigma^2 - \gamma_1(\gamma_0 - 1).$$

If we set $\gamma_0 = \gamma_1$, $c_0 = c_1$ this equation reduces to equation (25) for a "quasi-stationary" three-shock configuration. In the general case it is rather unmanageable. However, for an acoustic incident wave it is completely factorable and gives the physically likely solution

$$(36) \qquad \sigma'^2 = 1; \qquad \sin^2 \omega_r = (c_0/c_1)^2$$

and two additional solutions

$$(36') \qquad \begin{aligned} \sigma'^2 &= \frac{\gamma_0 - \gamma_1}{\gamma_0(\gamma_1 + 1)}; \qquad \csc^2 \omega_\tau = 0, \\ \sigma'^2 &= \frac{(c_1/c_0)^2(\gamma_1 - \gamma_0)}{\gamma_1(\gamma_0 + 1) - \gamma_0(c_1/c_0)^2(\gamma_1 + 1)}; \qquad \csc^2 \omega_\tau = \sigma'^2. \end{aligned}$$

It is possible to derive also a rather simple solution for the case of a weak incident wave, that is, for a shock wave which is almost acoustic. We set $\sigma^2 = 1 - \epsilon$, where ϵ is small so that ϵ^2 and terms of higher degree are negligible. We consider only the solution in the neighborhood of that given by equation (36). By substituting in equation (35) we obtain the interesting result

$$(37) \qquad \sigma'^2 = 1 + \epsilon \quad \text{and} \quad \sin^2 \omega_{r\epsilon} = \left(\frac{c_0}{c_1}\right)^2 \left[1 - \frac{2\gamma_0 + \gamma_1(\gamma_0 - 1)}{\gamma_1(\gamma_0 + 1)}\, \epsilon\right].$$

Hence for a weak shock there is always an angle $\omega_{r\epsilon}$ which produces total reflection whenever this is the case for an acoustic wave, and its value is given by equation (37).

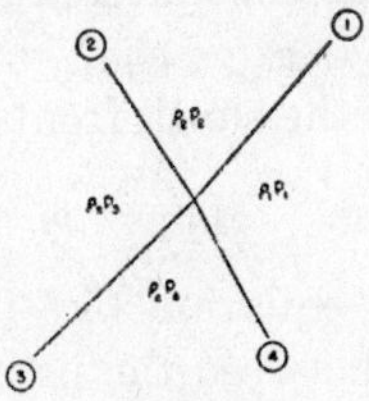

FIG. 15

7. Some remarks on four-shock configurations. We shall conclude this paper by deriving the condition under which a four-shock configuration may exist without a density discontinuity. We shall also indicate how this result may be extended to an n-shock configuration. It is known [7] that in the case of three-shock configurations a plane of density discontinuity must invariably accompany three intersecting shock waves. The following analogous theorem for four-shock configurations may be obtained by consideration of the pressure-density relation across a shock wave. Four intersecting shock waves cannot exist without the accompaniment of a plane of density discontinuity, except in two special cases. The first is obviously the case of two shock waves, symmetric to two other shock waves, with respect to a plane which replaces the wall in regular reflection. The second exception is the case where the pressure (or density) ratios across the first and third shocks, as well as across the second and fourth shocks, are equal. The shocks are taken in sequence around a circle.

The pressures and densities across a shock wave are related by the equation

$$\text{(5c)} \qquad \frac{p_i}{p_k} = \frac{\mu\rho_k - \rho_i}{\mu\rho_i - \rho_k}$$

where

$$\mu \equiv (\gamma - 1)/(\gamma + 1).$$

We assume that two-dimensional space is divided into four regions of uniform density and pressure by four intersecting shocks and apply equation (5c) to each of the four shocks (Fig. 15). We obtain the relation

$$\text{(38)} \qquad \frac{p_1}{p_2}\cdot\frac{p_2}{p_3} - \frac{p_1}{p_4}\cdot\frac{p_4}{p_3} = F(\mu) = (\mu\rho_2 - \rho_1)(\mu\rho_3 - \rho_2)(\mu\rho_4 - \rho_3)(\mu\rho_1 - \rho_4)$$
$$- (\mu\rho_1 - \rho_2)(\mu\rho_2 - \rho_3)(\mu\rho_3 - \rho_4)(\mu\rho_4 - \rho_1) = 0.$$

This expression is a third-degree polynomial in μ. It vanishes for $\mu = 0$, $\mu = 1$, $\mu = -1$; hence it cannot vanish also for $\mu = (\gamma - 1)/(\gamma + 1)$ ($(\gamma - 1)/(\gamma + 1)$ different from 0, 1, -1) unless it is identically zero. The coefficients of μ^3 and μ have the same factors

$$(\rho_2 - \rho_4)(\rho_1 - \rho_3)(\rho_1\rho_3 - \rho_2\rho_4);$$

while the coefficient of μ^2 is zero. Thus, $F(\mu) \equiv 0$ when any of the above factors vanish.

The condition $\rho_2 - \rho_4 = 0$ or $\rho_1 - \rho_3 = 0$ may be seen to be equivalent to the condition of symmetry of the type encountered in regular reflection from a rigid wall. If we set the third factor $(\rho_1\rho_3 - \rho_2\rho_4) = 0$ we obtain the following relations of the density ratios across the shock fronts

$$\rho_1/\rho_2 = \rho_4/\rho_3 \quad \text{or} \quad \rho_2/\rho_3 = \rho_1/\rho_4 \tag{39}$$

which must be satisfied. From equation (5c) it follows that this condition is equivalent to a similar relation between the pressure ratios

$$p_1/p_3 = p_4/p_3 \quad \text{or} \quad p_2/p_3 = p_1/p_4 . \tag{39'}$$

For an n-shock configuration equation (39) becomes

$$F(\mu) = (\mu - \eta_1)(\mu - \eta_2) \cdots (\mu - \eta_n) - (\mu - \eta_1')(\mu - \eta_2') \cdots (\mu - \eta_n') = 0, \tag{38'}$$

where

$$\eta_1 = \rho_1/\rho_2 , \quad \eta_2 = \rho_2/\rho_3 , \cdots , \quad \eta_{n-1} = \rho_{n-1}/\rho_n , \quad \eta_n = \rho_n/\rho_1 ,$$
$$\eta_1' = \eta_1^{-1}, \quad \eta_2' = \eta_2^{-1}, \cdots , \quad \eta_{n-1}' = \eta_{n-1}^{-1} , \quad \eta_n' = \eta_n^{-1} .$$

If we denote by $P_1 , P_2 , \cdots , P_n ; P_1' , P_2' , \cdots , P_n'$ the elementary symmetric functions of the η's and η''s, respectively, we may write equation (38′) in the form

$$F(\mu) = (\mu^n - P_1\mu^{n-1} + P_2\mu^{n-2} - \cdots + (-1)^{n-1}P_{n-1}\mu + (-1)^n P_n) - (\mu^n - P_1'\mu^{n-1} + P_2'\mu^{n-2} - \cdots + (-1)^{n-1}P_{n-1}'\mu + (-1)^n P_n') = 0. \tag{38''}$$

A necessary and sufficient condition for this equation to be identically zero is that the roots of the first polynomial on the right be equal to the roots of the second polynomial. But these roots are reciprocals of each other. Thus this condition is equivalent to the requirement that each of the roots of the first polynomial $\eta_1 , \eta_2 , \cdots , \eta_n$ be equal in some order to one of the reciprocals $\eta_1^{-1}, \eta_2^{-1} , \cdots , \eta_n^{-1}$. It will be noted that the product $P_n = \eta_1\eta_2 \cdots \eta_n = 1$.

For the case $n = 4$ this requirement is satisfied if

$$\begin{aligned} &(1)\ \eta_1\eta_2 = 1; \text{hence } \eta_3\eta_4 = 1,\\ &(2)\ \eta_1\eta_3 = 1; \text{hence } \eta_2\eta_4 = 1,\\ &(3)\ \eta_1\eta_4 = 1; \text{hence } \eta_2\eta_3 = 1. \end{aligned} \tag{39''}$$

These are, in turn, equivalent to the conditions derived above. We did not consider cases such as $\eta_1^2 = 1$, because this implies that the strength of shock (1) is zero, and hence reduces the problem to a three-shock configuration.

For any given value of n, there are also a number $(n - 1)$ of particular values of μ (or γ) for which a shock configuration without a density discontinuity may be obtained. To obtain these we note that the following relations hold between the elementary symmetric functions

$$(40)\quad P_1 = P'_{n-1}, \qquad P_2 = P'_{n-2}, \cdots, \qquad P_{n-1} = P'_1, \qquad P_n = P'_n = 1.$$

Hence equation (38″) reduces to

$$(39''')\quad \begin{aligned} -(P_1 - P_{n-1})\mu^{n-1} + (P_2 - P_{n-2})\mu^{n-2} - \cdots \\ + (-1)^{n-2}(P_{n-2} - P_2)\mu^2 + (-1)^{n-1}(P_{n-1} - P_1)\mu = 0. \end{aligned}$$

For $n = 3$ these values of μ are simply $0, -1$; while, for $n = 4$, $\mu = 0, -1, +1$.

To sum up, for $n = 3$, $n = 4$, physically plausible shock configurations without density discontinuities are possible only if equation (38″) is satisfied identically (that is, all coefficients are identically zero). Also, for all odd values of n (including $n = 3$) such configurations are possible for the physically unrealizable values of $\gamma = 0, 1$; while for even values of n (including $n = 4$) these are possible for $\gamma = 0, 1$, and ∞. For $n > 4$ equation (38″) has solutions other than the above, and hence configurations without density discontinuities may be expected to exist for any value of γ, without the above restrictions.

Bibliography

1. R. Becker, *Impact waves and detonation*, Zeitschrift für Physik vol. 8 (1922) p. 321.
2. H. A. Bethe, *The specific heat of air up to 25,000° C*, Division B, NDRC, 1942, OSRD Report No. 369.
3. H. A. Bethe and E. Teller, *Deviations from thermal equilibrium in shock waves*, Ballistic Research Laboratory, Aberdeen Proving Ground, 1940.
4. Stuart R. Brinkley, Jr., John G. Kirkwood, and John M. Richardson, *The Hugoniot curve for air from* 300–15,000° K, Division 8, NDRC, 1944, December 15 to January 15 Interim Report.
5. S. Chandrasekhar, *On the conditions for the existence of three-shock waves*, Ballistic Research Laboratory, Aberdeen Proving Ground, Report No. 367, 1943.
6. R. Courant and K. O. Friedrichs, *Interaction of shock and rarefaction waves in one-dimensional motion*, AMP Report 38.1R, 1943.
7. ———, *Supersonic flow and shock waves*, AMP Report 38.2R.
8. W. F. Durand, *Aerodynamic theory*, vol. 3, pp. 209–250.
9. K. Friedrichs, *Remarks on the Mach effect*, including Appendix, Division 8 and Applied Mathematics Panel, NDRC, 1943.
10. F. B. Harrison and W. Bleakney, *Remeasurement of reflection angles in regular and Mach reflection of shock waves*, Princeton University.
11. H. Hugoniot, *Sur la propagation du mouvement dans les corps et specialment dans les gaz parfaits*, Journal de l'Ecole Polytechnique vol. 58 (1889) p. 80.
12. P. C. Keenan and R. J. Seeger, *Analysis of data on shock intersections*, Progress Report I, Bureau of Ordnance, Explosives Research Report No. 15, 1944.
13. J. von Neumann, *Oblique reflection of shocks*, Bureau of Ordnance Explosives Research Report No. 12, 1943.
14. W. G. Penny and D. R. Davies, *Shock waves in air and their normal impact on plane rigid surfaces*, R.C. 214.
15. H. Polachek and R. J. Seeger, *Analysis of data on shock intersections*, Progress report II, NAVORD 417.

16. H. Polachek and R. J. Seeger, *Regular reflection of shocks in ideal gases*, Navy Department, Bureau of Ordnance ERR 13.

17. ———, *Regular reflection of shocks in water-like substances*, Bureau of Ordnance, Explosives Research Report No. 14, 1944.

18. J. W. Rayleigh, *Aerial plane waves of finite amplitude*, Proc. Roy. Soc. London vol. 84 (1910) pp. 274–284; Scientific Papers, vol. 5, pp. 573–610.

19. ———, *Theory of sound*, New York, 1945.

20. A. Schmidt, *Theory of the pressure impulse in gases and in the detonation wave*, Zeitschrift Gesellschaft für das gesamte Schiess und Sprengstoffwesser vol. 37 (1932) p. 145.

21. J. C. Smallwood, *Equations for specific heats of gases*, Industrial and Engineering Chemistry vol. 34 (1942) p. 863.

22. Lincoln G. Smith, *Photographic investigation of the reflection of plane shocks in air*, OSRD No. 6271.

23. A. H. Taub, *Refraction of plane shock waves*, Physical Review vol. 72 (1947) pp. 51–60.

24. G. I. Taylor and J. W. Maccoll, *The mechanics of compressible fluids*.

25. L. H. Thomas, *Note on Becker's theory of the shock front*, Journal of Chemical Physics vol. 12 (1944) p. 449.

26. H. Weyl, *A scheme for the computation of shock waves in gases and fluids*, Applied Mathematics Panel, NDRC, Memorandum 38.7, 1943.

Naval Ordnance Laboratory,
Washington, D. C.

THE BREAKING OF WAVES IN SHALLOW WATER

BY

J. J. STOKER

The subject of hydrodynamics furnishes material particularly suitable for a conference on non-linear problems in mechanics of continua since these problems are practically all non-linear in character. In particular, problems of wave motion in water with a free surface are non-linear—because of the free surface conditions—even where a velocity potential exists. In this talk we consider the special case of motions in shallow water; these non-linear problems should be of peculiar interest at the present conference because they can be formulated and discussed in terms of the same basic mathematical theory that is used for the discussion of one-dimensional non-steady flows of a compressible gas. The problems concerned with compressible gas dynamics have, quite naturally in view of the prevailing interests due to the recent war, occupied most of the attention at this conference. It may therefore be interesting and amusing to see how the underlying mathematical theory developed for gas dynamics can be interpreted in terms of the phenomena of surface gravity waves in water.

It has already been stated that we are interested in surface waves in *shallow* water. By this we mean that the depth is small compared with the length of the surface waves. The theory used is then an approximation which consists essentially in a development of all quantities in the exact hydrodynamical theory with respect to the depth, and neglect of all terms beyond a certain order in the depth.[1] If the lowest order terms only are retained it turns out that the pressure in the water is, as in hydrostatics, simply proportional to the depth below the free surface, while the velocity of the water is independent of the depth coordinate. In case the original undisturbed depth of the water is constant the differential equations take the form, in the case of a two-dimensional flow:

$$h(u_t + uu_x) = -\bar{p}_x, \tag{1}$$

$$(hu)_x + h_t = 0, \tag{2}$$

in which $h(x, t)$ is the height of the surface above the bottom, $u(x, t)$ is the velocity in the x-direction, and $\bar{p}$ is the quantity

$$\bar{p} = \frac{g}{2} h^2. \tag{3}$$

One observes that the equations (1) and (2) are exactly the same as in gas dynamics if the depth of the water h is interpreted as the "density" and the quantity

[1] For a complete treatment of the subject of this talk, including an extensive bibliography, see Stoker, *The formation of breakers and bores: The theory of nonlinear wave propagation in shallow water and open channels*; Communications of the Institute of Mathematics and Mechanics, New York University, January 1948. The shallow water theory itself was derived long ago by Lagrange.

$\bar{p}$ as the "pressure"; in addition, these two quantities obey the "adiabatic relation" (3) with the fixed adiabatic exponent 2.

The problem of calculating the manner in which a disturbance created at a point in shallow water propagates into still water of constant depth is exactly analogous to the problem of calculating how a pressure wave travels down a tube filled with a uniform gas initially at rest. The latter problem has been studied in great detail both theoretically and experimentally. Perhaps the most striking single result is the following: there is a very great and fundamental difference in the propagation of a rarefaction wave in a gas as compared with the propagation of a compression wave. A rarefaction wave propagates indefinitely as a smooth, continuous motion, but a compression wave always develops into a discontinuous shock wave no matter how smoothly the compression wave may have been started. The latter phenomenon, which is very well known experimentally, mirrors itself in the mathematical treatment of the problem, since it is readily shown that a continuous compression wave solution cannot exist for an indefinite time but that rather a discontinuity must eventually occur. In water waves the analogous situation is the following: a depression wave (that is, a wave created by steadily lowering the water level at a certain point) propagates indefinitely into still water of uniform depth as a continuous wave, but the creation of an elevation or hump in the water always leads to a discontinuity. *The central theme of the present talk is that the occurrence of breakers in shallow water is to be interpreted as the analogue of the occurrence of discontinuous shock waves in gas dynamics.*

In discussing the propagation of a surface wave into still water the well known method of characteristics is particularly useful since it brings out the essential phenomena mentioned above very clearly. One can also calculate quite easily by this means the time and distance to the point at which a discontinuity first develops, which turns out to be in all cases the point at which the slope of the water surface first becomes infinite. Beyond this point, which we call somewhat arbitrarily the *breaking point*, the theory embodied in equations (1), (2), (3) is no longer adequate to describe the motion; numerical calculations of the shape of the wave as it progresses into still water indicate strongly, however, that the wave would actually curl over and break very soon after passing this point.

After a wave curls over and breaks it is often observed that a steady motion ensues in which a steep and turbulent front progresses into still water at constant speed leaving water of constant and greater depth than in front behind it. Such a wave is called a bore. In rivers and estuaries in various parts of the world from England to China such bores are a common occurrence as the result of the incoming tides. They are the exact analogues of a steady shock wave passing down a tube filled with gas at rest. The theory of such bores is, as one would expect, analogous to the corresponding theory of discontinuous shock waves in gases, with one exception: only mechanical sources of energy are considered in the water wave theory and this brings with it (as Rayleigh first noticed) a violation of the energy law for water particles crossing the shock. In gas dynamics the energy balance is upheld on crossing the shock—only the entropy increases discontinuously; in the case of a bore we can only say that the mechanical energy lost is used to establish a turbulent regime behind the shock.

Thus we are able to treat the change in form of waves up to the breaking point (in the sense defined above) and we can also study bores or shock waves which under appropriate circumstances are the end result of the breaking of a smooth wave. But there is an intermediate phase which starts at the point we have called (somewhat arbitrarily, perhaps) the breaking point and continues on through a period in which the wave curls over and finally falls into the still water in front to cause a very turbulent motion. If the water behind the wave should then continue to move forward at nearly constant speed the result would be a bore. It is not entirely out of the question to treat this intermediate phase in the development of a bore, and this is a problem which should be studied.

The theory of long waves in shallow water embodied in equations (1), (2), and (3) is a simplified version, or perhaps better, a special case of the theory used by hydraulic engineers to discuss unsteady flows in open channels. In particular, the theoretical treatment of the problem of propagation of flood waves in rivers requires the solution of equations of the same general type as (1) and (2), but more complicated because of the necessity to consider a sloping bottom and to take account of the variable cross section of the river channel. Basically, however, the method of characteristics can still be applied to obtain numerical solutions. The French hydraulic engineers, following a tradition which goes back to Boussinesq, have carried out studies of this kind in various special cases.[2] This type of problem deserves much further study, both because of its practical importance and because it offers scope for developments of mathematical interest.

New York University,
New York N. Y.

[2] The nature of some of this work is indicated in the paper of the speaker cited above.

ON HAMILTON'S PRINCIPLE FOR PERFECT COMPRESSIBLE FLUIDS[1]

BY

A. H. TAUB

1. Introduction. It is the purpose of this paper to show the equivalence of a variational problem which we shall call Hamilton's principle with the equations describing the conservation of momentum and energy of a perfect compressible fluid, one with no viscosity or heat conductivity. The equations we refer to are those holding in regions where the flow is continuous and those holding across discontinuities. The classical form of Hamilton's principle for a continuous medium is

$$\int_{t_1}^{t_2} (\delta L + \delta A)\, dt = 0$$

where δA is the virtual work done by the displacement against the external forces and the Lagrangean L is of the form of a volume integral over the portion of the medium being considered,

$$L = \int \mathfrak{L}\, d\tau.$$

The Lagrangean density $\mathfrak{L}$ is in turn a function of generalised coordinates $q^i(x, t)$ and their derivatives with respect to the variables indicated, x and t, which refer to space and time variables.

L. Lichtenstein[2] has given a discussion of Hamilton's principle for fluids which is incomplete in that the conservation of energy is not considered. His treatment differs from that given below in that he does not consider variations in L produced by variations in the temperature distribution in the fluid and hence does not obtain one Euler condition from his variational problem. The discussion given here will parallel his, but the Lagrangean will be generalised and the type of variations considered will be extended.

E. Lamla[3] has derived a relativistic isentropic theory from a variational principle. This theory does not consider discontinuities. The Lagrangean used below may be modified so as to give a relativistically invariant theory of non-isentropic hydrodynamics. These results will be presented elsewhere.

2. Notation and definitions. Two cartesian sets of coordinates will be used

---

[1] This work was done in part while the author was employed at Princeton University on work sponsored by the Office of Naval Research, contract N6ori—105 Task II.

[2] L. Lichtenstein, *Grundlagen der Hydrodynamik*, Berlin, Springer, 1929, Chap. 9.

[3] E. Lamla, *Über die Hydrodynamik des Relitivitatsprinzips*, Dissertation, Berlin, 1912, and Annalen der Physik (4) vol. 37 (1912) p. 772.

below, the Lagrange coordinates denoted by x_0^i $(i = 1, 2, 3)$ and the Euler coordinates denoted by x^i $(i = 1, 2, 3)$. Thus

$$x^i = x^i(X_0, t) \tag{2.1}$$

where

$$x^i(X_0, 0) = x_0^i$$

are the equations of the particle paths of the fluid. The components of velocity will be written as

$$u^i(x_0, t) = \frac{dx^i}{dt} = \left(\frac{\partial x^i}{\partial t}\right)_{x_0^i}.$$

The square of the velocity will be written as

$$u^2 = u^i u^i$$

where we have used the summation convention which will be used throughout the paper. The velocity components may also be considered as functions of x^i through the inverses of equations (2.1).

The symbol d/dt will be used for the partial derivative of a function with respect to time where the x_0^i are kept constant and the symbol $\partial/\partial t$ for that in which the x^i are kept constant. Thus

$$\frac{d}{dt} f(x, t) = \frac{\partial f}{\partial t} + u^i \frac{\partial f}{\partial x^i}.$$

Let V_0, V and V_1 denote the volumes occupied by the fluid under consideration at times t_0, t, and t_1 respectively where t is some time such that $t_0 \leqq t \leqq t_1$. Let S_0, S, and S_1 be the respective boundaries of these volumes. Part of these boundaries, denoted by S_0', S', and S_1', may consist of rigid walls moving with prescribed velocities and the remainder, denoted by S_0'', S'', S_1'', consists of free boundaries. We shall assume that each S is made up of a finite number of analytic and regular surfaces.

The functions x^i and u^i will be assumed to be continuous single-valued functions of x_0^i and t in and on S in the interval $t_0 \leqq t \leqq t_1$. However we shall allow du^i/dt to have discontinuities across a surface $\Sigma(t)$ which varies with time within V. For simplicity we shall assume that a single such surface exists and that it divides V into two parts V^1 and V^2. For each value of t, $\Sigma(t)$ will also be assumed to be made up of a finite number of analytic and regular parts.

We shall write for the varied paths

$$x^{*i} = x^i + e\xi^i \tag{2.2}$$

and for the varied velocities

$$u^{*i} = u^i + e\frac{d\xi^i}{dt} \tag{2.3}$$

where ξ^i and $d\xi^i/dt$ are assumed to be continuous functions of x_0^i and t for x_0^i in V_0 and on S_0 and $t_0 \leqq t \leqq t_1$.

We shall denote by λ_i the components of the inward drawn normal to a surface bounding a region. We shall assume that

$$\xi^i(x_0, t_0) = \xi^i(x_0, t_1) = 0 \qquad \text{and} \qquad \lambda_i \xi^i = 0 \tag{2.4}$$

on S'. Thus the variations to be considered will be those which have no component normal to the moving rigid walls S'. If S' has an edge we shall take the ξ^i in the direction of the edge and if S' has a corner we shall assume that ξ^i vanishes there.

The components of external body forces per unit mass will be denoted by X_i and of the surface forces per unit area by $X_{\sigma i}$. These will be assumed to be continuous functions in V and on S. The forces $X_{\sigma i}$ will be assumed to vanish on S'.

Let $\rho_0(x_0)$ be the density of the fluid at the time $t = t_0$. This will be assumed to be a piecewise differentiable function of x_0^i in V_0 and on S_0. The density of the fluid at any later time will be denoted by ρ, which will be considered to be a function of x^i and t.

The continuity of mass is then insured by requiring that

$$\frac{\rho_0}{\rho} = J = \det \left\| \frac{\partial x^i}{\partial x_0^i} \right\| \tag{2.5}$$

Thus we obtain

$$\int_{V_0} \rho_0 \, d\tau_0 = \int_V \rho \, d\tau.$$

From equation (2.5) and the rule for differentiating a determinant it follows that

$$-\frac{1}{\rho}\frac{d\rho}{dt} = \frac{\partial u^i}{\partial x^i}. \tag{2.6}$$

If the paths are varied then the density at time t will be changed from ρ to ρ^* where

$$\frac{\rho_0}{\rho^*} = \det \left\| \frac{\partial x^{*i}}{\partial x_0^i} \right\| = \frac{\rho_0}{\rho} \det \left\| \frac{\partial x^{*i}}{\partial x^i} \right\|. \tag{2.7}$$

From this equation we may readily compute that

$$-\left(\frac{\rho}{\rho^{*2}}\frac{d\rho}{de}\right)_{e=0} = \frac{\partial \xi^i}{\partial x^i} = -\frac{1}{\rho}\left(\frac{d\rho^*}{de}\right)_{e=0} \tag{2.8}$$

where of course ρ^* at $e = 0$ is ρ. Hence for arbitrary volumes

$$\int_{V^*} \rho^* \, d\tau^* = \int_V \rho \, d\tau = \int_{V_0} \rho_0 \, d\tau_0,$$

where V_0, V, and V^* are the volumes occupied by any given amount of fluid at time t_0, at time t where the particle paths are given by (2.1), and at the time t when the particle paths are given by (2.2) respectively.

The internal energy per unit mass of the gas will be denoted by U and it is assumed this is a known function of the density ρ and the absolute temperature T. The entropy per unit mass will be denoted by S and we have from the definition of entropy and the law of conservation of energy the relation

$$T\,dS = dU + pd(1/\rho). \tag{2.9}$$

We denote the Helmholtz function per unit mass, sometimes called the free energy per unit mass, by H where

$$H = U - TS$$

and H will be considered as a function of ρ and T. It follows from (2.9) that

$$\left(\frac{\partial H}{\partial \rho}\right)_T = \frac{1}{\rho^2}\,p \tag{2.10}$$

and

$$\left(\frac{\partial H}{\partial T}\right)_\rho = -S. \tag{2.11}$$

Variations in the temperature distribution of the gas will be considered below. However these will not be produced by variations in $T(x, t)$ directly but by considering the temperature as the time derivative of another variable introduced by Helmholtz[4] and called a by Von Laue.[5] Thus

$$T = da/dt. \tag{2.12}$$

We shall consider variations of this variable and write them as

$$\begin{gathered} a^* = a + e\alpha, \\ T^* = \frac{da}{dt} + e\,\frac{d\alpha}{dt} = T + e\,\frac{d\alpha}{dt}, \\ \alpha(x_0\,, t_0) = \alpha(x_0\,, t_1) = 0 \end{gathered} \tag{2.13}$$

where a and α are functions of x_0^i and t or alternatively of x^i, t, which are assumed to be piecewise continuous and to have piecewise continuous derivatives. The function α will be required to vanish on the boundary S. Across the surface of discontinuity T will not be continuous as is to be expected from the usual formulation of shock wave theory.

The moving surface of discontinuity, Σ, will be assumed to have either of the alternative parametric forms

$$x_0^i = x_0^i(u, v, t) \quad \text{or} \quad x^i = x^i(u, v, t). \tag{2.14}$$

---

[4] H. Von Helmholtz, *Wissenschaftliche Abhandlungen*, Leipzig, Barth, 1921, vol. 1, p. 248.
[5] M. Von Laue, *Relativitätstheorie*, Braunschweig, Vieweg, 1921, vol. 1, p. 248.

The "varied" surface Σ^* will then be expressed as

(2.15) $$x_0^{*i} = x_0^i(u, v, t) + e\Xi_0^i(u, v, t)$$

or as

(2.16) $$x^{*i} = x^i(u, v, t) + e\Xi^i(u, v, t).$$

We shall assume that the surfaces Σ and Σ^* respectively divide the volumes V and V^* into subvolumes V_1 and V_2 and V_1^* and V_2^*.

It will be necessary to consider the derivative with respect to e of integrals of the form

$$I(e, t) = \int_{V_1^*+V_2^*} \rho^* f(x^*)\, d\tau^* = \int_{V_1^*} \rho^* f\, d\tau^* + \int_{V_2^*} \rho^* f\, d\tau^*$$

where the volumes V_1^* and V_2^* are volumes occupied at time t by particles of the fluid which at time $t = t_0$ were in volumes $V_{01}(e)$ and $V_{02}(e)$ respectively, that is the particles have Lagrange coordinates belonging to $V_{01}(e)$ and $V_{02}(e)$ respectively. The paths of these particles are given by (2.2) and the discontinuity surface $\Sigma(e, t)$ is given by (2.15). Thus

(2.17) $$I(e, t) = \int_{V_{01}(e)} \rho_0 f(x^*(x_0, t, e))\, d\tau_0 + \int_{V_{02}(e)} \rho_0 f(x^*(x_0, t, e))\, d\tau_0 .$$

Note that

$$I(0, t) = \int_{V_{01}} \rho_0 f(x(x_0, t))\, d\tau_0 + \int_{V_{02}} \rho_0 f(x(x_0, t))\, d\tau_0$$

where the subvolumes V_{01} and V_{02} are those separated by the discontinuity $\Sigma = \Sigma(0, t)$ given by the first of (2.14) and the particle paths are given by (2.1).

It follows from (2.17) and the fact that $\xi^i(x^0, t_0) = 0$ that

(2.18) $$\frac{\partial I(e, t)}{\partial e} = \int_{V_{01}+V_{02}} \rho_0 \frac{\partial f}{\partial e}\, d\tau_0 + \int_{\Sigma(e)} \rho_0 \Xi^i \lambda_i [f]\, d\sigma$$

and

(2.19) $$\left(\frac{\partial I}{\partial e}\right)_{e=0} = \int_{V_{01}+V_{02}} \rho_0 \frac{\partial f}{\partial e}\, d\tau_0 + \int_{\Sigma(0)} \rho_0 \Xi^i \lambda_i [f]\, d\sigma$$

where

$$[f] = f_2 - f_1 .$$

f_2 and f_1 are the values of f on the V_2 and V_1 sides of $\Sigma(e)$ respectively, λ^i are the components of the normal of $\Sigma(e)$ directed from the region $V_2(e)$ toward the region $V_1(e)$ and $d\sigma$ is the surface element on $\Sigma(e)$.

The following formula will also be needed below

(2.20) $$\frac{d}{dt}\int_{V_1(t)+V_2(t)} \rho f(x, t)\, d\tau = \frac{d}{dt}\int_{V_{01}+V_{02}} \rho_0 f(x(x_0, t), t)\, d\tau_0$$
$$= \int_{V_{01}+V_{02}} \rho_0 \frac{df}{dt}\, d\tau_0 + \int_{\Sigma} \rho_0 V_0^i \lambda_i [f]\, d\sigma$$

where V_0^i are the cartesian components of velocity of the discontinuity surface Σ relative to Lagrange coordinates.

The quantity $\rho_0 V_0^i \lambda_i$ represents the rate at which matter is crossing the discontinuity from region V_{02} to region V_{01}. If ρ_1, u_1^i, and ρ_2, u_2^i represent the density and particle velocity on the two sides of the discontinuity Σ and if V^i are the cartesian components of the velocity of the discontinuity in Euler coordinates then since mass is to be conserved in crossing the discontinuity we must have

$$\rho_0 V_0^i \lambda_i = \rho_1 (V^i - u_1^i)\lambda_i = \rho_2 (V^i - u_2^i)\lambda_i . \tag{2.21}$$

3. Hamilton's principle. The Lagrangean we shall use for the variational problem is

$$L = \int_{V_1+V_2} \rho(2^{-1}u^2 - H(\rho, T))\, d\tau.$$

This differs from that of Lichtenstein in that we employ the function H instead of any function which satisfies (2.10). Our procedure will also differ from that of Lichtenstein in that we shall allow variations of T, through the variable a, and also variations in the surface of discontinuity.

Following the usual technique of the calculus of variations we consider the integral

$$I(e) = \int_{t_0}^{t_1} \left\{ \int_{V_1^*+V_2^*} \rho^*[2^{-1}u^{*2} - H(\rho^*, T^*) + e\xi^i X_i]\, d\tau^* + e \int_{S''*} X_{\sigma i}\xi^i\, d\sigma^* \right\} dt \tag{3.1}$$

where u^{*i} is given by (2.3), ρ^* by (2.7), T^* by (2.13), the boundary between the subvolumes V_1^* and V_2^* is the discontinuity Σ^* which is given in parametric form by (2.15) or (2.16), and $d\tau^*$ and $d\sigma^*$ are the volume elements in V^* and on S''^* respectively.

Hamilton's principle is then embodied in the statement that the particle paths, temperature distribution and motion of the surface of discontinuity is such that

$$I'(0) = \left(\frac{dI}{de}\right)_{e=0} = 0. \tag{3.2}$$

Using (2.18), (2.10), (2.13), (2.7) and (2.8) we obtain

$$I'(0) = \int_{t_1}^{t_2} \left\{ \int_{V_1+V_2} \left[\rho u^i \frac{d\xi^i}{dt} + p\frac{\partial \xi^i}{\partial x^i} + \rho X_i \xi^i + \rho S \frac{d\alpha}{dt} \right] d\tau + \int_{S''} \xi^i X_{\sigma i}\, d\sigma + \int_\Sigma \rho_0 \Xi_0^i [2^{-1}u^2 - H]\, d\sigma \right\} dt. \tag{3.3}$$

We now note that

$$\int_{V_1+V_2} \rho \left(u^i \frac{d\xi^i}{dt} + S \frac{d\alpha}{dt} \right) d\tau = \int_{V_{01}+V_{02}} \rho_0 \left(u^i \frac{d\xi^i}{dt} + S \frac{d\alpha}{dt} \right) d\tau_0$$

$$= \int_{V_{01}+V_{02}} \rho_0 \frac{d}{dt} (u^i \xi^i + S\alpha)\, d\tau_0 - \int_{V_{01}+V_{02}} \rho_0 \left(\frac{du^i}{dt} \xi^i + \frac{dS}{dt} \alpha \right) d\tau_0$$

$$= \frac{d}{dt} \int_{V_{01}+V_{02}} \rho_0 (u^i \xi^i + S\alpha)\, d\tau_0 - \int_\Sigma \rho_1 (V^i - u_1^i) \lambda_i [u^i \xi^i + S\alpha]\, d\sigma$$

$$- \int_{V_{01}+V_{02}} \rho_0 \left(\frac{du^i}{dt} \xi^i + \frac{dS}{dt} \alpha \right) d\tau_0 ,$$

where we have made use of (2.20) and (2.21). Moreover

$$\int_{V_1+V_2} p \frac{\partial \xi^i}{\partial x^i}\, d\tau = \int_{V_1+V_2} \frac{\partial}{\partial x^i} (p\xi^i) - \int_{V_1+V_2} \frac{\partial p}{\partial x^i} \xi^i\, d\tau$$

$$= -\int_{S''} p\xi^i \lambda_i\, d\sigma + \int_\Sigma [p] \xi^i \lambda_i\, d\sigma - \int_{V_1+V_2} \frac{\partial p}{\partial x^i} \xi^i\, d\tau$$

where in the first surface integral λ_i are the components of the inward drawn normal on S'' and in the second integral λ_i is defined as in equation (2.19). In this equation we have made use of the last of equations (2.4).

In virtue of these results and the vanishing of ξ^i and α at times t_0 and t_1 equation (3.3) becomes

$$I'(0) = -\int_{t_1}^{t_2} \left\{ \int_{V_1+V_2} \left(\left(\rho \frac{du^i}{dt} + \frac{\partial p}{\partial x^i} - \rho X_i \right) \xi^i + \rho \frac{dS}{dt} \alpha \right) d\tau \right.$$

$$+ \int_{S''} \xi^i (X_{\sigma_i} - p\lambda_i)\, d\sigma$$

(3.4)

$$+ \int_\Sigma \{ [p]\lambda_i - \rho_1 (V^i - u_1^i) \lambda_i [u^i]) \xi^i$$

$$\left. + \rho_0\, \Xi_0^i [2^{-1} u^2 - H] - \rho_1 (V^i - u_1^i) \lambda_i [S\alpha] \} \, d\sigma \right\} dt.$$

If we now require that $I'(0) = 0$ for arbitrary ξ^i and α subject to the conditions (2.4) and (2.13) we must have[6]

$$\rho \frac{du^i}{dt} + \frac{\partial p}{\partial x^i} = \rho X_i , \qquad \frac{dS}{dt} = 0 \tag{3.5}$$

in V_1 and in V_2. On S'' we must have

$$X_{\sigma_i} = p\lambda_i \tag{3.6}$$

[6] Lichtenstein, loc. cit. p. 367.

and across Σ we must have

$$\rho_1(V^j - u_1^j)\lambda_j[u^i] = [p]\lambda_i \tag{3.7}$$

and

$$\rho_0\Xi_0^i\lambda_i[2^{-1}u^2 - H] = \rho_1(V^i - u_1^i)\lambda_i[S\alpha]. \tag{3.8}$$

The first of equations (3.5) are the usual equations of motion of a compressible fluid and (3.7) expresses the conservation of momentum across the surface of discontinuity. It is well known that for continuous flows the conservation of energy is equivalent to the second of (3.5). We shall now show that equation (3.8) may be used to establish a relation between $[S\alpha]$ and $\Xi_0^i\lambda_i$ such that energy is conserved across Σ.

4. **The energy integral.** It is well known in particle mechanics that the constant of motion called the energy may be obtained from the Lagrangean function as follows

$$\mathcal{E} = \sum \dot{q}^i \frac{\partial \mathfrak{L}}{\partial \dot{q}^i} - \mathfrak{L}$$

where q^i are the generalised coordinates and $\dot{q}^i$ are the velocities corresponding to these coordinates. This suggests considering the expression

$$\int_{V_1+V_2} \rho\mathcal{E}\, d\tau$$

and seeing under what conditions its time derivative vanishes as a consequence of the equations of motion where in our case

$$\mathcal{E} = u^i \frac{\partial}{\partial u^i}(2^{-1}u^2 - H(\rho, T)) - T\frac{\partial H}{\partial T} - (2^{-1}u^2 - H) = 2^{-1}u^2 + U.$$

This is of course the total energy per unit mass of the fluid. Now

$$\frac{d}{dt}\int_{V_1+V_2} \rho\mathcal{E}\, d\tau = \frac{d}{dt}\int_{V_{01}+V_{02}} \rho_0\,\mathcal{E}(x_0\,, t)\, d\tau_0$$

$$= \int_{V_1+V_2} \rho\frac{d\mathcal{E}}{dt}\, d\tau + \int_\Sigma \rho_1(V^i - u_1^i)\lambda_i[\mathcal{E}]\, d\sigma.$$

But

$$\int_{V_1+V_2} \rho\frac{d\mathcal{E}}{dt} = \int_{V_1+V_2} \rho\left(u^i\frac{du^i}{dt} + T\frac{dS}{dt} + \frac{p}{\rho^2}\frac{d\rho}{dt}\right) d\tau$$

where we have used equation (2.9). If we now use equations (3.5) and (2.6) we obtain

$$\int_{V_1+V_2} \rho\frac{d\mathcal{E}}{dt}\, d\tau = \int_{V_1+V_2} \left(\rho X_i u^i - \frac{\partial(pu^i)}{\partial x^i}\right) d\tau$$

which may be written as

$$\int_{V_1+V_2} \rho \frac{d\mathcal{E}}{dt}\, d\tau = \int_{V_1+V_2} \rho X_i u^i + \int_S p u^i \lambda_i \, d\sigma - \int_\Sigma [p u^i]\lambda_i \, d\sigma.$$

Hence

$$\int_{V_1+V_2} \rho \frac{d\mathcal{E}}{dt}\, d\tau = \int_{V_1+V_2} \rho X_i u^i \, d\tau + \int_{S''} X_{\sigma_i} u^i \, d\sigma + \int_{S'} p u^i \lambda_i \, d\sigma - \int_\Sigma [p u^i]\lambda_i \, d\sigma.$$

The first two integrals in these expressions give the work done on the fluid by the external forces. The third gives the work done by the rigid walls moving against the pressure of the fluid.

The conservation of energy will be a consequence of the variational principle, that is, we shall have

$$\frac{d}{dt}\int_{V_1+V_2} \rho\mathcal{E}\, d\tau = \int_{V_1+V_2} \rho X_i u^i \, d\tau + \int_{S''} X_{\sigma_i} u^i \, d\sigma + \int_{S'} p u^i \lambda_i \, d\sigma \tag{4.1}$$

if we relate $[S\alpha]$ to $\rho_0 \Xi_0^i \lambda_i$ so that

$$[p u^i]\lambda_i = \rho_1 (V^i - u_1^i)\lambda_i [2^{-1} u^2 + U] \tag{4.2}$$

is a consequence of (3.8). This may readily be done as follows: Write

$$Y = \frac{\rho_0 \Xi^i \lambda_i}{\rho_1 (V^i - u_1^i)\lambda_i} \tag{4.3}$$

and set

$$[S\alpha] = Y\left[ST + u^2 - \frac{p u^i \lambda_i}{\rho_1 (V^i - u_1^i)\lambda_i}\right]. \tag{4.4}$$

Then (4.2) is a consequence of (3.8). Thus if the variation of the discontinuity surface is not allowed to be arbitrary but is determined in terms of α, which may be taken to be continuous across Σ, by (4.3) of the discontinuity and (4.4) then conservation of energy throughout the fluid is assured.

Hence we have shown that the conservation of momentum and energy in regions of continuity and across discontinuities are equivalent to the condition $I(0) = 0$ where the particle paths, the temperature (thought of as the velocity of a) and the position of the discontinuity are varied. The variations are restricted in that they have been required to satisfy certain conditions: (1) The variations in the particle paths are assumed to vanish at times t_0 and t_1 and to be normal to the moving rigid walls. (2) The variation in the variable a is also assumed to vanish at t_0 and t_1 and moreover if a discontinuity is present its variation must be related to that of a through equations (4.3) and (4.4).

Contrary to the situation in particle dynamics, if discontinuities are present, the conservation of energy is not a consequence of the Euler equations of variation but is achieved by using a relation between the variation in the discontinuity and the variable a.

UNIVERSITY OF WASHINGTON,
SEATTLE, WASH.

THE FOUNDATIONS OF THE THEORY OF ELASTICITY

BY

F. D. MURNAGHAN

1. **Introduction.** The theory of elasticity deals with the strain, or deformation, which a deformable medium experiences when the medium is subjected to stress. In the usual form of the theory the strain is measured from the unstressed state of the medium and it is assumed that the strain is infinitesimal, that is, that only the lowest order terms in the expansion of any function of the strain components in a series of powers of these components need be considered. The relation connecting stress and strain (which relation is known as Hooke's Law) is found from energy considerations. Confining our attention to isothermal processes it is assumed that the work done on the medium by the stress applied to it is stored up in the medium in the form of free energy and that the density of this energy (that is, the energy per unit volume) is a function of the strain-components. Assuming that the strain is measured from the state of zero stress the lowest power terms in the expansion of the energy density in a series of powers of the strain components are of the second order and Hooke's Law appears in its well known form:

Each component of stress is the derivative of the volume density of the free (or elastic) energy with respect to the corresponding strain component so that each stress component is a linear function of the strain components, or, equivalently,

The stress tensor is the gradient of the volume density of the elastic energy with respect to the strain tensor and the stress tensor is a linear function of the strain tensor.

This simple result (on which the whole theory of strength of materials is based) is in accord with experience only for very small strains. For example in the stretching of an iron bar Hooke's Law stating the proportionality of stress and strain may be in accord with experience only for extensions which are not greater than one-tenth of one per cent while the bar may undergo, without rupturing, an extension of 30%. The problem of large extensions of a bar is complicated by the phenomenon of plastic flow and we shall devote our attention here mainly to the simpler problem of compression of a medium under extreme hydrostatic compression, in which problem plastic flow does not appear. This simpler problem can be treated without introducing the concepts of strain and stress components since the strain is completely determined by the change in volume and the stress by the pressure, but we shall not follow this procedure since the theory we propose has wider application than that of compressibility under hydrostatic pressure. When we apply the theory to this problem of compressibility and make certain simplifying hypotheses suggested by the results of experiment we obtain the following formula connecting pressure with volume:

$$p = A\left\{\left(\frac{v_0}{v}\right)^a - 1\right\}$$

where v_0 is the volume when $p = 0$ and A and a are constants which depend on the medium. Professor Bridgman published in 1945 observations on compressibility up to the high pressure of 100,000 kilograms per square centimeter. We select his observations on indium for comparison with the formula just given. Bridgman gives the values of v/v_0 at $p = 10{,}000, 20{,}000, \cdots$ up to 100,000 kg/cm^2 (1 kilogram per square centimeter being, approximately, atmospheric pressure). For $p = 10{,}000$ kg/cm^2, $v/v_0 = .9760$ and for $p = 100{,}000$ kg/cm^2, $v/v_0 = .8451$ so that at the highest pressure the medium has been compressed a little more than 15 %. On assigning to the constants A and a in our formula the values $A = 97{,}750$; $a = 4.2$ we obtain the following values of p at the values of v/v_0 which are given by Bridgman:

10 080, 19 970, 30 100, 39 800, 49 680, 59 640, 69 640, 79 750, 89 980, 100 200.
The greatest difference between the observed and calculated values is less than 1% while the experimental errors are estimated to be as much as 3%. If our formula is valid up to a compression of as much as 50% we find that a pressure of 1.7×10^6 kg/cm^2 is necessary to compress indium to half its volume.

Before proceeding to show how our formula connecting pressure and volume was derived, we point out a feature of the formula which is interesting:

The formula treats pressure and hydrostatic tension (that is, negative pressure) unsymmetrically. Thus p must be infinite to make v zero while p is $-A$, and not $-\infty$, when $v = \infty$. In other words the medium will rupture at a hydrostatic tension $\leqq A$. In particular indium cannot support a hydrostatic tension of 100,000 atmospheres while a hydrostatic pressure of 100,000 atmospheres compresses it by less than 16% of its original volume. All theories of rupture hitherto proposed which have come to our attention have failed to meet this crucial test of differentiating between hydrostatic pressure and hydrostatic tension and so are seriously inadequate; for a medium cannot be crushed by any amount of hydrostatic pressure while it can be torn apart by sufficiently high hydrostatic tension.

2. Outline of elasticity theory. We denote by A and P the positions of a typical particle of our deformable medium when the medium is unstrained and strained, respectively. Let a be the 1×3 matrix whose elements are the coordinates (a^1, a^2, a^3) of A with respect to any convenient Cartesian reference frame:

$$a = \begin{pmatrix} a^1 \\ a^2 \\ a^3 \end{pmatrix}.$$

Then the elements of the 3×1 matrix da are the coordinates of the vector element of arc for the unstrained medium and the squared element of arc for the unstrained medium is given by

$$ds_a^2 = da^* \, da$$

where da^* is the 1×3 matrix obtained by transposing da. Similarly if x is the 3×1 matrix whose elements are the coordinates of P with respect to any convenient Cartesian reference frame (not necessarily the same as that adopted for the unstrained medium) the squared element of arc for the strained medium is given by

$$ds_x^2 = dx^* \, dx.$$

Let us denote by J the 3×3 Jacobian matrix of the coordinates (x^1, x^2, x^3) with respect to the coordinates (a^1, a^2, a^3), that is, the derivative of the matrix x with respect to the matrix a:

$$J = \partial x / \partial a.$$

(Note that the variables x tell the rows, and the variables a the columns, of J.) Then

$$dx = J \, da; \qquad dx^* = da^* J^*$$

so that

$$ds_x^2 = da^* J^* \, J da.$$

Hence

$$ds_x^2 - ds_a^2 = da^*(J^*J - E_3) \, da$$

where E_3 is the unit 3×3 matrix. The strain, or deformation, is said to be measured by the difference between the two matrices J^*J and E_3 and one-half this difference is termed the strain matrix or strain tensor. On denoting the (symmetric) strain matrix by ϵ we have

$$\epsilon = 2^{-1}(J^*J - E_3); \qquad J^*J = E_3 + 2\epsilon.$$

The elements of ϵ are invariants (that is, scalars) under transformation of the rectangular Cartesian coordinates (x^1, x^2, x^3) of P; on the other hand ϵ is a tensor of order two under transformation of the rectangular Cartesian coordinates (a^1, a^2, a^3) of A.

In order to express the conditions of equilibrium of the strained medium we introduce the concept of a virtual deformation of the medium. This is done by regarding the matrix x as a function not only of a but of an accessory variable or parameter. We denote differentiation with respect to this parameter by the symbol δ and observe that since the elements of a and the parameter are independent the order in which differentiations with respect to any element of a and the parameter are performed is immaterial. Thus

$$\delta J = \delta\left(\frac{\partial x}{\partial a}\right) = \frac{\partial}{\partial a}(\delta x) = \frac{\partial}{\partial x}(\delta x) J$$

and so

$$\delta J^* = J^* \left\{ \frac{\partial}{\partial x} (\delta x) \right\}^*,$$

$$\delta\epsilon = 2^{-1}\{(\delta J^*)J + J^*\delta J\} = J^*DJ$$

where D is the symmetric 3×3 matrix:

$$D = 2^{-1}\left[\left\{\frac{\partial}{\partial x}(\delta x)\right\}^* + \frac{\partial}{\partial x}(\delta x)\right].$$

A virtual deformation is said to be *rigid* if $\delta\epsilon = 0$ or, equivalently, if $D = 0$ and the condition of equilibrium of our strained medium may be phrased as follows:

The strained medium is in equilibrium if, and only if, the virtual work of the forces acting on any portion of the medium in any rigid virtual deformation is zero.

We denote by T the 3×3 matrix whose elements are the coordinates of the stress tensor. T is a function of the matrix x and it is a tensor of the second order under transformations of the rectangular Cartesian coordinates (x^1, x^2, x^3) of P. On writing out the virtual work we find that a linear form in the elements of the matrices δx and $\partial(\delta x)/\partial x$ must be zero for any rigid virtual deformation. On considering the virtual translations (for which $\partial(\delta x)/\partial x$ is the zero matrix) we obtain the relation div $T + \rho(x)F = 0$ where $\rho(x)$ is the density of the strained medium and F is the 1×3 matrix whose elements are the coordinates of the applied force per unit mass; div T is the 3×1 matrix obtained by taking, in turn, the divergence of each column of T as if these columns were vectors. The relation div $T + \rho(x)F = 0$ eliminates the elements of δx from the expression for the virtual work and so we have a linear form in the elements of the matrix $\partial(\delta x)/\partial x$ which must be zero for every rigid virtual deformation, that is, for every skew-symmetric matrix $\partial(\delta x)/\partial x$. The coefficients of this linear form are the elements of T and so we find (as part of the conditions for equilibrium) that T must be a symmetric matrix. This being the case our linear form in the elements of $\partial(\delta x)/\partial x$ may be written as a linear form in the elements of $D = 2^{-1}[\{\partial(\delta x)/\partial x\}^* + \partial(\delta x)/\partial x]$ and so our conditions for equilibrium, namely

$$\text{div } T + \rho(x)F = 0; \qquad T^* = T,$$

are not only necessary but are also sufficient. These conditions of equilibrium being satisfied the virtual work of the forces acting on any portion $V(x)$ of the medium in *any* virtual deformation is the integral over $V(x)$ of the sum of the diagonal elements of (that is, the trace of) the product TD. This trace of TD (which we shall denote by $[TD]$) is the sum of the products of each element of T by the corresponding element of D. Thus

$$\text{Virtual work} = \int_{V(x)} [TD]\, dV(x).$$

On denoting by ϕ the free energy per unit mass of the medium, the total free energy stored up in $V(x)$ is $\int_{V(x)} \rho(x)\phi \, dV(x)$ and so the principle of energy conservation yields

$$\int_{V(x)} [TD] \, dV(x) = \delta \int_{V(x)} \rho(x)\phi \, dV(x).$$

By virtue of the principle of conservation of mass the right-hand side of this equation may be written as

$$\delta \int_{V(a)} \rho(a)\phi \, dV(a) = \int_{V(a)} \rho(a)\delta\phi \, dV(a) = \int_{V(x)} \rho(x)\delta\phi \, dV(x)$$

and so

$$\int_{V(x)} [TD] \, dV(x) = \int_{V(x)} \rho(x)\delta\phi \, dV(x).$$

Since this relation must hold for *every* portion $V(x)$ of the strained medium we have

$$[TD] = \rho(x)\delta\phi.$$

So far we have been on sure ground; the fundamental relation we have just written is merely a convenient form of statement of the principle of conservation of energy. To proceed further we must make some hypotheses concerning $\delta\phi$. The theory of elasticity assumes that ϕ is a function of the strain matrix ϵ and so $\delta\phi$ is the sum of the diagonal elements (that is, the trace) of the product of $\delta\epsilon$ by the gradient $\partial\phi/\partial\epsilon$ of ϕ with respect to ϵ. This hypothesis is not *necessary* and its adoption rules out any possibility of an explanation of the phenomenon of hysteresis. A more general hypothesis would be that $\delta\phi$ is a linear form in the elements of $\delta\epsilon$ whose coefficients might well depend on T as well as on ϵ. This linear form might well be a non-exact differential and if this were the case ϕ would depend not only on the strain ϵ but also on the "curve" connecting the unstrained and strained states of the medium, that is, ϕ depends not only on the final strained position of the medium but on the "history" of the medium during its passage from the unstrained state to the strained state. Furthermore in order to explain phenomena which arise after the plastic state has been reached, $\delta\phi$ must treat in an unsymmetrical way "loading" and "unloading"; in other words the stresses observed when strain is being relieved are different from the stresses experienced when strain is being imposed on the medium.

If we commit ourselves to the hypothesis that ϕ is a function of ϵ it is clear that, due to the fact that ϵ is a symmetric matrix, ϕ may be written in many equivalent ways. We *normalise* ϕ by replacing, if necessary, each element of ϵ by the corresponding element of $2^{-1}(\epsilon^* + \epsilon)$; in other words we write ϕ as a symmetrical function of ϵ. The essential advantage of this procedure is that $\partial\phi/\partial\epsilon$

becomes a symmetric matrix. On using the relation $\delta\epsilon = J^*DJ$ our fundamental formula $[TD] = \rho(x)\delta\phi$ yields

$$[TD] = \rho(x)\left[\frac{\partial\phi}{\partial\epsilon}\,\delta\epsilon\right] = \rho(x)\left[\frac{\partial\phi}{\partial\epsilon}\,J^*DJ\right]$$

and since the trace of the product of two matrices is insensitive to the order in which the factor matrices are taken this is equivalent to

$$[TD] = \rho(x)\left[J\,\frac{\partial\phi}{\partial\epsilon}\,J^*D\right].$$

Since this must hold for an arbitrary virtual deformation, that is, for an arbitrary symmetric matrix D, we must have

$$T = \rho(x)J\,\frac{\partial\phi}{\partial\epsilon}\,J^*.$$

This is the fundamental formula which reduces, in the classical (or infinitesimal) theory of elasticity to Hooke's Law:

$$T = \rho(a)\,\frac{\partial\phi}{\partial\epsilon} = \frac{\partial}{\partial\epsilon}\,\{\rho(a)\phi\}.$$

To see how this reduction takes place we observe that, by virtue of the principle of conservation of mass, $\rho(x)\,dV(x) = \rho(a)\,dV(a)$ so that

$$\begin{aligned}\rho(x) = \rho(a)\,|\det J\,|^{-1} &= \rho(a)\{\det\,(J^*J)\}^{-1/2} = \rho(a)\{\det\,(E + 2\epsilon)\}^{-1/2}\\ &= \rho(a)\{1 - I_1 + \cdots\}\end{aligned}$$

where I_1 is the trace, or sum of the diagonal elements, of the strain matrix ϵ. The infinitesimal theory of elasticity assumes that the difference between $J = \partial x/\partial a$ and the unit matrix E_3 is an infinitesimal matrix. If we denote the matrix $x - a$ by U we have $J = E_3 + \partial u/\partial a$, $J^* = E_3 + (\partial u/\partial a)^*$. In accordance, then, with the basic principle of the infinitesimal theory, namely, that only the lowest order terms in the expansion of any function as a power series in the elements of the infinitesimal matrix $\partial u/\partial a$ need be considered, we have

$$J = E_3; J^* = E_3; \epsilon = 2^{-1}(J^*J - E_3) = 2^{-1}\left\{\left(\frac{\partial u}{\partial a}\right)^* + \frac{\partial u}{\partial a}\right\};\quad \rho(x) = \rho(a)$$

and so

$$T = \rho(a)\,\frac{\partial\phi}{\partial\epsilon} = \frac{\partial}{\partial\epsilon}\,\{\rho(a)\phi\}.$$

It must be emphasized that this relation is only valid in the infinitesimal theory; in the exact theory (or in a closer approximation to the exact theory than that furnished by the infinitesimal theory) T, while it is a function of the matrix $\partial u/\partial a$ (which matrix is not, in general, symmetric), is not, in general, a function of the symmetric matrix $\epsilon = 2^{-1}(J^*J - E_3)$. It is incorrect to follow the procedure which has been adopted by some authors who have tried to construct a theory of *finite* deformations. They propose to keep terms other than those of the lowest order in the expressions for the strain components, so that they use the exact expression $\epsilon = 2^{-1}(J^*J - E_3)$ rather than the infinitesimal approximation $\epsilon = 2^{-1}\{(\partial u/\partial a)^* + \partial u/\partial a\}$ but they also propose to use the formula $T = \partial\{\rho(a)\phi\}/\partial\epsilon$. If we wish to improve the infinitesimal theory the only logical thing to do is to make a clear-cut second approximation in accordance with which we keep in each function of $\partial u/\partial a$ terms whose order is higher by one than that of the lowest order terms. Thus

$$J = E_3 + \frac{\partial u}{\partial a}; \qquad J^* = E_3 + \left(\frac{\partial u}{\partial a}\right)^*; \qquad \epsilon = 2^{-1}(J^*J - E_3);$$

$$\rho(x) = \rho(a)(1 - I_1); \quad T = \rho(a)(1 - I_1)\left(E_3 + \frac{\partial u}{\partial a}\right)\frac{\partial \phi}{\partial \epsilon}\left\{E_3 + \left(\frac{\partial u}{\partial a}\right)^*\right\}.$$

If we denote by ϕ_1 the terms of the lowest order in ϕ (after the zero order terms) and by ϕ_2 the terms in ϕ whose order is greater by one than that of the terms of ϕ_1 we obtain

$$\begin{aligned} T &= \rho(a)(1 - I_1)\left(E_3 + \frac{\partial u}{\partial a}\right)\left(\frac{\partial \phi_1}{\partial \epsilon} + \frac{\partial \phi_2}{\partial \epsilon}\right)\left\{E_3 + \left(\frac{\partial u}{\partial a}\right)^*\right\} \\ &= \frac{\partial}{\partial \epsilon}\{\rho(a)\phi_1\} + \frac{\partial}{\partial \epsilon}\{\rho(a)\phi_2\} - I_1\frac{\partial}{\partial \epsilon}\{\rho(a)\phi_1\} + \frac{\partial u}{\partial a}\frac{\partial}{\partial \epsilon}\{\rho(a)\phi_1\} \\ &\qquad + \frac{\partial}{\partial \epsilon}\{\rho(a)\phi_1\}\left(\frac{\partial u}{\partial a}\right)^*. \end{aligned}$$

The stress in the unstrained position is furnished by the formula

$$T_0 = \frac{\partial}{\partial \epsilon}\{\rho(a)\phi_1\}_{\epsilon=0}$$

of the infinitesimal theory and so this stress will be zero if, and only if, the terms of ϕ_1 are of degree greater than 1 in the elements of the strain matrix ϵ.

A deformable medium is said to be *elastically isotropic* when ϕ is insensitive to a rotation of the rectangular Cartesian reference frame with respect to which the co-ordinates of A are furnished by the elements of the matrix a. It follows that for an isotropic elastic medium, ϕ is a function of the three invariants I_1, I_2 and I_3 of the strain matrix ϵ. Here I_1 is the sum of the diagonal elements of ϵ; I_2 is the sum of the three two-rowed diagonal minors of ϵ and I_3 is the determinant of ϵ. Thus I_1 is of the first, I_2 of the second, and I_3 of the third degree in the elements of ϵ. If ϕ is developed as a power series in I_1, I_2 and I_3 the terms in ϕ which are of the first degree in the elements of ϵ are a multiple of I_1. Hence the stress in the unstrained position must be, if it is not zero, a hydrostatic pressure or tension (for the gradient of I_1 with respect to ϵ is the unit matrix). This result, which is of fundamental importance, may be phrased as follows:

If a deformable medium is subjected to a stress in the unstrained state which is not a hydrostatic pressure or tension the medium is not elastically isotropic.

3. **The theory of scalar strain and stress.** We term a strain or stress *scalar* when its matrix is a scalar matrix, that is, a multiple of the unit matrix. Thus a stress is scalar when it is either a hydrostatic pressure or a hydrostatic tension. We consider the relation connecting a scalar strain with a scalar stress and we write

$$T = -pE_3; \qquad \epsilon = -eE_3$$

(so that p is the hydrostatic pressure and e is connected with the density by means of the formula $\rho(x) = \rho(a)(1 - 2e)^{-3/2)}$. ϕ, being a function of ϵ, is a function of e and, hence, is a function of $\rho(x)$ so that $\partial\phi/\partial\epsilon = (\partial\phi/\partial\rho(x))(\partial\rho(x)/\partial\epsilon)$. From the basic relation $\rho(x) = \rho(a)\{\det(E + 2\epsilon)\}^{-1/2}$ we obtain, by logarithmic differentiation,

$$\frac{1}{\rho(x)}\frac{\partial\rho(x)}{\partial\epsilon} = -(E + 2\epsilon)^{-1}$$

and so

$$\frac{\partial\phi}{\partial\epsilon} = -\rho(x)\frac{\partial\phi}{\partial\rho(x)}(E + 2\epsilon)^{-1}.$$

On taking the trace of the matrices on each side of the fundamental equation

$$-pE_3 = \rho(x)J\frac{\partial\phi}{\partial\epsilon}J^*$$

we obtain

$$-3p = \rho(x)\left[J\frac{\partial\phi}{\partial\epsilon}J^*\right] = \rho(x)\left[\frac{\partial\phi}{\partial\epsilon}J^*J\right]$$
$$= \rho(x)\left[\frac{\partial\phi}{\partial\epsilon}(E + 2\epsilon)\right] = -3\rho^2(x)\frac{\partial\phi}{\partial\rho(x)}$$

and so

$$p = \rho^2(x)\frac{\partial\phi}{\partial\rho(x)}.$$

We suppose the medium to be homogeneous, that is, that $\rho(a)$ is independent of a and that e is constant over it; then $\rho(x)$ is constant and $\rho(x)v = m$ where v is the volume of the (strained) medium and m its mass. m being constant we have

$$\frac{\partial\varphi}{\partial\rho(x)} = \frac{-m}{\rho^2(x)}\frac{\partial\phi}{\partial v}$$

and so

$$p = -m \frac{\partial \phi}{\partial v} = -\frac{\partial \Phi}{\partial v}$$

where $\Phi = m\phi$ is the elastic energy stored up in the deformed medium. This simple formula replaces, for scalar strain and stress, the formula $T = \rho(x)J(\partial\phi/\partial\epsilon)J^*$ which is valid for any type of strain.

Our whole attention must now center on the following question:

What is the nature of the function Φ of v?

The usual way of answering this question is to regard Φ as expanded in a power series in $v/v_0 - 1$ where v_0 is the volume in the unstrained state. The coefficient of the first power of $v/v_0 - 1$ in this expansion must be $-p_0v_0$ where p_0 is the pressure in the unstrained state. If the unstrained state is unstressed, that is, if $p_0 = 0$, the linear theory yields

$$p = -C\left(\frac{v}{v_0} - 1\right) = -C\frac{\Delta v}{v_0}$$

where C is a constant. The next approximation would be

$$p = -C\frac{\Delta v}{v_0} + \frac{D}{2}\left(\frac{\Delta v}{v_0}\right)^2$$

where D is a second constant, and so on. This procedure has the nature of an "action at a distance" theory. It tries to bridge the strain from A to P by piling up more and more detailed information about what is happening at A; thus C is the product by $-v_0$ of the value of dp/dv at $v = v_0$, D is the product by v_0^2 of the value of d^2p/dv^2 at $v = v_0$, and so on. We propose to replace this action at a distance theory (which always looks back to the distant state of zero stress) by a "local" or "action through a medium" theory. Thus we regard the strain from A to P as built up from a succession of infinitesimal strains by the method of integration. We consider, then, the pressure $p + \Delta p$ when the volume is $v + \Delta v$ where now the unstrained position is that for which the volume is v and the pressure p. The increment in Φ is of the form

$$\Delta\Phi = -p\Delta v + \frac{q}{2}(\Delta v)^2 + \cdots$$

and

$$p + \Delta p = -\frac{\partial(\Phi + \Delta\Phi)}{\partial(v + \Delta v)} = -\frac{\partial(\Delta\Phi)}{\partial(\Delta v)} = p - q\Delta v + \cdots$$

so that

$$dp/dv = -q.$$

In order to obtain some information about q we follow the procedure of the infinitesimal theory of elasticity and set

$$\rho\Delta\phi = -pI_1 + ce^2 + \cdots$$

(note that ρ takes the place of $\rho(a)$; $\rho(x)$ is now $\rho + \Delta\rho$). It is not necessary for us to assume our deformable medium to be isotropic; however if it is isotropic the second order terms in $\rho\Delta\phi$ are $((\lambda + 2\mu)/2)I_1^2 - 2\mu I_2$ where (λ, μ) are Lamé's familiar elastic constants. Since $I_1 = -3e$, $I_2 = 3e^2$ we have $c = (3/2)(3\lambda + 2\mu)$. Now

$$\frac{v + \Delta v}{v} = \frac{\rho}{\rho + \Delta\rho} = (1 - 2e)^{3/2}$$

and so

$$e = \frac{1}{2}\left\{1 - \left(1 + \frac{\Delta v}{v}\right)^{2/3}\right\} = -\frac{1}{3}\frac{\Delta v}{v} + \frac{1}{18}\left(\frac{\Delta v}{v}\right)^2 + \cdots$$

Hence

$$I_1 = \frac{\Delta v}{v} - \frac{1}{6}\left(\frac{\Delta v}{v}\right)^2 + \cdots$$

so that

$$\rho\Delta\phi = -p\left(\frac{\Delta v}{v}\right) + \left(\frac{1}{9}\,c + \frac{1}{6}\,p\right)\left(\frac{\Delta v}{v}\right)^2 + \cdots$$

Since $\Phi = m\phi = \rho v\phi$ it follows that

$$\Delta\Phi = -p\Delta v + \left(\frac{1}{9}\,c + \frac{1}{6}\,p\right)\frac{(\Delta v)^2}{v} + \cdots$$

and so $vq = 2c/9 + p/3 = (\lambda + 2\mu/3) + p/3$.

The quantity $vq = -v\,dp/dv$ is the reciprocal of the compressibility, that is, the *incompressibility* of the medium. Denoting it by K we have the following result: *For an isotropic elastic medium the incompressibility K is furnished by the formula*

$$K = (\lambda + 2\mu/3) + p/3,$$

where λ, μ are the Lamé elastic constants of the medium.

We cannot make further progress without some information as to the nature of the dependence of $\lambda + 2\mu/3$ on the pressure p in the unstrained state. From the experiments on indium referred to in the introduction we find that K may be approximated very well by a linear function of p over the very wide range of pressure extending from $p = 0$ to $p = 100{,}000$ kg/cm^2. On writing $K = a(p + A)$ where a and A are constants we have to integrate the equation

$$-v\,\frac{dp}{dv} = a(p + A).$$

We obtain

$$p = A\left\{\left(\frac{v_0}{v}\right)^a - 1\right\}$$

where v_0 is the value of v when $p = 0$. For indium a has the value 4.2. a is the value of dK/dp and our approximating assumption is that dK/dp is constant. In interpreting the results of experiments on compressibility the expression $K_0 = -v_0\, dp/dv$ is frequently used rather than K. Since $K = aA\,(v_0/v)^a$, $K_0 = aA\,(v_0/v)^{a+1}$ and so

$$\frac{dK_0}{dp} = -\frac{v_0}{K_0}\frac{dK_0}{dv} = (a+1)\frac{v_0}{v}.$$

From an examination of the results of experiment Slater[1] has estimated that for metals the value of dK_0/dp at $p = 0$, that is, $a + 1$, is approximately 5. Our theory which is based on the assumption that dK/dp (not dK_0/dp) is constant yields the value 5.2 for the value of dK_0/dp at $p = 0$ for indium.

If we do not wish to make the assumption that K is a linear function of p (or, equivalently, that $\lambda + 2\mu/3$, the medium being assumed elastically isotropic, is a linear function of p) we have to consider the equation

$$-v\, dp/dv = (\lambda + 2\mu/3) + p/3$$

where λ and $2\mu/3$ is furnished by experiment (or by assumptions concerning the nature of the force between particles of the medium) as a function of p. If we approximate $(\lambda + 2\mu/3) + p/2$ by a quadratic, instead of a linear, function of p we obtain the formula

$$\frac{v_0}{v} = \left(1 + \frac{p}{A}\right)^{1/a}\left(1 + \frac{p}{B}\right)^{1/b}, \qquad A,\ a,\ B,\ b,\ \text{constants},$$

instead of the previous formula

$$\frac{v_0}{v} = \left(1 + \frac{p}{A}\right)^{1/a}.$$

If we wish to focus our attention on $\lambda + 2\mu/3$ rather than on $(\lambda + 2\mu/3) + p/3$ we should introduce the variable $u = (v/v_0)^{1/3}$; then $v\, dp/dv + p/3 = (u dp/du + p)/3 = (1/3)\, d(pu)/du$ and the equation we have to solve is

$$\frac{d}{du}(pu) = -(3\lambda + 2\mu).$$

From this point of view the most convenient form of statement of the law connecting p and v is

$$p = \left(\frac{v_0}{v}\right)^{1/3}\left\{f\left[\left(\frac{v_0}{v}\right)^{1/3}\right] - f(1)\right\}$$

[1] J. C. Slater, *Note on Grüneisen's constant for the incompressible metals*, Physical Review vol. 57 (1940) pp. 744–746.

where f is a function which must be determined by experiment or by making assumptions concerning the forces between elementary particles of the medium.

4. **The uniform tension of an elastic cylinder.** We consider an elastic cylinder (whose generators we take to be parallel to the c-axis) and suppose this deformed by forces applied to the top and base, the rest of the surface of the cylinder being free from applied force. Our problem is a boundary-value problem: The stress-matrix must not only satisfy the conditions of equilibrium

$$\operatorname{div} T = 0; \qquad T^* = T$$

but also the force applied to any element of area of the surface of the strained medium must have its coordinates furnished by the elements of the 1×3 matrix $dS(x)T$. Here $dS(x)$ is the 1×3 matrix

$$dS(x) = (d(x^2, x^3), d(x^3, x^1), d(x^1, x^2)).$$

This *boundary condition* reveals a difficulty which lies at the very core of the subject; we do not know, until the problem is solved, the boundary of the *strained* medium but, nevertheless, we must determine T so that $dS(x)T$ furnishes the coordinates of the force applied to the (unknown) boundary $S(x)$ of the medium. We do know the surface $S(a)$ of the unstrained medium and so we try to express our boundary condition in terms of the element of area $dS(a)$ of $S(a)$. $dS(a)$ is the 1×3 matrix

$$dS(a) = (d(a^2, a^3), d(a^3, a^1), d(a^1, a^2))$$

and it is easy to see that

$$\rho(x)\, dS(x) = \rho(a)\, dS(a)J^{-1}.$$

In fact

$$\begin{aligned} d(x^2, x^3) &= \frac{\partial(x^2, x^3)}{\partial(a^2, a^3)}\, d(a^2, a^3) + \frac{\partial(x^2, x^3)}{\partial(a^3, a^1)}\, d(a^3, a^1) + \frac{\partial(x^2, x^3)}{\partial(a^1, a^2)}\, d(a^1, a^2) \\ &= \left| \frac{\partial(x^1, x^2, x^3)}{\partial(a^1, a^2, a^3)} \right| \left\{ d(a^2, a^3)\, \frac{\partial a^1}{\partial x^1} + d(a^3, a^1)\, \frac{\partial a^2}{\partial x^1} + d(a^1, a^2)\, \frac{\partial a^3}{\partial x^1} \right\} \end{aligned}$$

and since $| \partial(x)/\partial(a) | = \rho(a)/\rho(x)$ it follows that $dS(x) = (\rho(a)/\rho(x))\, dS(a)J^{-1}$. Since $T = \rho(x)J(\partial\phi/\partial\epsilon)J^*$ it follows that

$$dS(x)T = \rho(a)\, dS(a)\, \frac{\partial\varphi}{\partial\epsilon}\, J^*.$$

We shall denote by $T(a)$ the 3×3 matrix $\rho(a)\partial\varphi/\partial\epsilon = \partial\{\rho(a)\phi\}/\partial\epsilon$ so that $T(a)$ is the stress matrix of the classical infinitesimal theory and then our boundary condition may be framed as follows:

The force applied to the element of area of the strained medium has its coordinates furnished by the elements of the 1×3 matrix $dS(a)T(a)J^*$.

The conditions of equilibrium may readily be written in a form which brings into prominence the 3×3 matrix $T(a)J^*$ (rather than $T = (\rho(x)/\rho(a))JT(a)J^*$). We have only to write the virtual work in any virtual deformation, namely,

$$\int_{S(x)} dS(x)\, T\delta x + \int_{V(x)} \rho(x) F\delta x\, dV(x)$$

in the equivalent form

$$\int_{S(a)} dS(a)\, T(a)J^*\delta x + \int_{V(a)} \rho(a) F\delta x\, dV(a)$$

and then to write down that this virtual work is zero for a virtual translation. We find

$$\text{div } (a)\{T(a)J^*\} + \rho(a)F = 0.$$

Here div (a) is the 1×3 matrix differential operator

$$\text{div } (a) = \left(\frac{\partial}{\partial a^1}, \frac{\partial}{\partial a^2}, \frac{\partial}{\partial a^3}\right).$$

We shall consider only the case where all applied forces are surface forces; then F is the zero 1×3 matrix and the conditions of equilibrium appear in the form

$$\text{div } (a)\{T(a)J^*\} = 0.$$

In the classical infinitesimal theory J^* is replaced by the unit matrix E_3 and the conditions of equilibrium are written in the form div $(a)T(a) = 0$. This procedure is exact in the case of a *homogeneous strain,* that is, a strain in which the strain-matrix ϵ is a constant function of a. In fact it is not difficult to show that when $J^*J = E_3 + 2\epsilon$ is a constant function of a so also is J (the only hypothesis necessary for this being the existence and continuity of the second derivatives of the variables x with respect to the variables a). When J is a constant matrix the matrix div $(a)\{T(a)J^*\}$ reduces to $\{\text{div } (a)T(a)\}J^*$ and since J^* is nonsingular the equation div $(a)\{T(a)J^*\} = 0$ reduces to div $(a)T(a) = 0$. In the problem with which we are concerned (namely, the *uniform* tension of an elastic cylinder) the strain is homogeneous so that the conditions of equilibrium reduce to div $(a)T(a) = 0$. Since $T(a) = \partial\{\rho(a)\phi\}/\partial\epsilon$ and since ϵ is constant these equations are automatically satisfied if $\rho(a)$ is constant, that is, if the cylinder is uniformly dense. We assume that this is the case and the question that confronts us is the following:

What is the nature of the constant matrix $T(a)$?

To answer this question we observe that since the sides of the cylinder are free from applied force the matrix $T(a)J^*$ must have its first two rows zero (for the force on any element $dS(x)$ of the strained cylinder is furnished by the 1×3 matrix $dS(a)T(a)J^*$). Hence the first two rows of $T(a)$ must be zero and since $T(a)$

is a symmetric matrix the only element of $T(a)$ which is different from zero is the element in the third row and third column. Denoting this element by P we have

$$T(a) = \begin{pmatrix} 0 & 0 & 0 \\ 0 & 0 & 0 \\ 0 & 0 & P \end{pmatrix}.$$

The case treated by the classical theory is that in which the medium is initially unstressed and is elastically isotropic. Then $\rho(a)\phi = ((\lambda + 2\mu)/2)I_1^2 - 2\mu I_2$ where λ, μ are Lamé's elastic constants. Thus $T(a) = \lambda I_1 E_3 + 2\mu\epsilon$ so that the constant strain matrix ϵ is diagonal. Thus $J^*J = E + 2\epsilon$ is diagonal and there is consequently no lack of generality in taking J to be diagonal (since J is the product of a diagonal matrix by a rotation matrix). Denoting the diagonal elements of ϵ by $(\epsilon_1, \epsilon_2, \epsilon_3)$ we find, since the diagonal elements of $T(a)$ are $(0,0,P)$, that $\epsilon_1 = \epsilon_2 = -\sigma\epsilon_3$ where $\sigma = \lambda/2(\lambda + \mu)$ is Poisson's ratio and that $\epsilon_3 = P/E$ where $E = \mu(3\lambda + 2\mu)/(\lambda + \mu)$ is Young's modulus. To take care of the case where the medium is subjected to a stress of the type

$$T_0 = \begin{pmatrix} 0 & 0 & 0 \\ 0 & 0 & 0 \\ 0 & 0 & P_0 \end{pmatrix}$$

when the strain is zero we must write $\rho(a)\phi$ in the form $\rho(a)\phi_1 + \rho(a)\phi_2$ where $\rho(a)\phi_1 = P_0\epsilon_{cc}$ and $\rho(a)\phi_2$ is a quadratic function of the elements of the strain matrix ϵ. Since we know that the medium is not elastically isotropic we cannot write $\rho(a)\phi_2$ in the familiar form $((\lambda + 2\mu)/2)I_1^2 - 2\mu I_2$. If the medium is isotropic when free from stress it will be elastically insensitive to rotation around the c-axis when subjected to the stress T_0. Then $\rho(a)\phi_2$ must be of the form

$$\rho(a)\phi_2 = \{(\lambda + 2\mu)I_1^2 - 4\mu I_2 + \alpha\epsilon_{cc}^2 + 4\beta(\epsilon_{aa}\epsilon_{bb} - \epsilon_{ab}\epsilon_{ba}) + 2\gamma\epsilon_{cc}(\epsilon_{aa} + \epsilon_{bb})\}/2$$

where α, β and γ are three elastic constants which reduce to zero when P_0 is zero. Thus (α, β, γ) are functions of P_0 as are also (λ, μ). The expression for $T(a)$ is

$$T(a) = \begin{pmatrix} 0 & 0 & 0 \\ 0 & 0 & 0 \\ 0 & 0 & P_0 \end{pmatrix} + \lambda I_1 E_3 + 2\mu\epsilon + \alpha \begin{pmatrix} 0 & 0 & 0 \\ 0 & 0 & 0 \\ 0 & 0 & \epsilon_{cc} \end{pmatrix}$$

$$+ 2\beta \begin{pmatrix} \epsilon_{bb} & -\epsilon_{ab} & 0 \\ -\epsilon_{ba} & \epsilon_{aa} & 0 \\ 0 & 0 & 0 \end{pmatrix} + \gamma \begin{pmatrix} \epsilon_{cc} & 0 & 0 \\ 0 & \epsilon_{cc} & 0 \\ 0 & 0 & \epsilon_{aa} + \epsilon_{bb} \end{pmatrix}$$

and since

$$T(a) = \begin{pmatrix} 0 & 0 & 0 \\ 0 & 0 & 0 \\ 0 & 0 & P \end{pmatrix}$$

it follows that ϵ is a diagonal matrix (for $\mu - \beta \neq 0$ since β reduces to zero when $P_0 = 0$). On denoting the diagonal elements of ϵ by $\epsilon_1, \epsilon_2, \epsilon_3$ we obtain $\epsilon_1 = \epsilon_2 = -\sigma\epsilon_3$ where σ (Poisson's ratio) is furnished by the formula

$$\sigma = \frac{\lambda + \gamma}{2(\lambda + \mu + \beta)}.$$

Furthermore

$$P - P_0 = \{\lambda + 2\mu + \alpha - 2\sigma(\lambda + \gamma)\}\epsilon_3.$$

The stress matrix T is furnished by the formula appropriate for media which are under stress when the strain is zero:

$$T = T(a) - I_1 \frac{\partial}{\partial \epsilon} \{\rho(a)\phi_1\} + \frac{\partial u}{\partial a} \frac{\partial}{\partial \epsilon} \{\rho(a)\phi_1\} + \frac{\partial}{\partial \epsilon} \{\rho(a)\phi_1\} \left(\frac{\partial u}{\partial a}\right)^*.$$

Since ϵ is diagonal we may assume, without any loss of generality, that $\partial u/\partial a$ is a diagonal matrix whose diagonal elements are $(\epsilon_1, \epsilon_2, \epsilon_3)$. It follows that all elements of T are zero save that in the third row and third column which element has the value $P - P_0 I_1 + 2P_0\epsilon_3 = P + P_0\epsilon_3(1 + 2\sigma)$. Thus

$$T_{33} = P_0 + \epsilon_3\{P_0(1 + 2\sigma) + \lambda + 2\mu + \alpha - 2\sigma(\lambda + \gamma)\}.$$

The total applied force is the product of the unstrained cross-sectional area A by $P(1 + \epsilon_3)$ (since J^* is the diagonal matrix whose elements are $(1 + \epsilon_1, 1 + \epsilon_2, 1 + \epsilon_3)$). Hence Young's modulus (the quotient of this total applied force by $A\epsilon_3$) is given by the formula

$$E = \lambda + 2\mu + \alpha - 2\sigma(\lambda + \gamma) + P_0.$$

We are now ready to obtain the formula connecting stress and strain by the process of integrating a succession of infinitesimal homogeneous strains. To do this we consider the strain from $P:(x, y, z)$ to $P + \Delta P:(x + \Delta x, y + \Delta y, z + \Delta z)$ (we give up the superscript notation since our theory is now over). We have $(z + \Delta z) = (1 + \epsilon_3)z$ so that $\epsilon_3 = \Delta z/z$ and so

$$\Delta T_{zz} = \frac{\Delta z}{z} \{T_{zz}(1 + 2\sigma) + \lambda + 2\mu + \alpha - 2\sigma(\lambda + \gamma)\}.$$

If l is the length of the cylinder when $T_{zz} = P_0$ it follows, on integration, that

$$\log \frac{z}{l} = \int_{P_0}^{T_{zz}} \frac{dT_{zz}}{(1 + 2\sigma)T_{zz} + (\lambda + 2\mu + \alpha) - 2\sigma(\lambda + \gamma)}.$$

In order to carry out the indicated integration we must make some hypothesis (preferably guided by experimental evidence) concerning the nature of the dependance of the elastic constants λ, μ, α, β, γ upon T_{zz}. It is particularly important to know how $1 + 2\sigma$ (the coefficient of T_{zz} in the denominator of the integrand) depends on T_{zz}. In the theory of hydrostatic pressure we have seen that the assumption that $\lambda + 2\mu/3$ is a linear function of p gave results in accord with experiment over a very wide range of pressure. If we assume that $1 + 2\sigma$, $\lambda + 2\mu + \alpha$ and $2\sigma(\lambda + \gamma)$ may be approximated closely enough by linear functions of T_{zz} the denominator of our integrand is a quadratic function of T_{zz} whose constant term is the value of $(\lambda + 2\mu + \alpha) - 2\sigma(\lambda + \gamma)$ at $T_{zz} = 0$ (P_0 being taken to be zero). Since we have assumed the medium to be elastically isotropic when free from stress this is the value of $\lambda + 2\mu - 2\sigma\lambda$, that is, of Young's modulus E, at zero stress. If σ decreases as T_{zz} increases our quadratic function of T_{zz} has two real zeros of opposite sign; as T_{zz} approaches the positive one of these $z \to \infty$ and as T_{zz} approaches the negative zero $z \to 0$. The relation connecting $T_{zz} = T$ and z is of the form

$$\frac{1 + T/A}{1 - T/B} = \left(\frac{z}{l}\right)^a, \qquad A, B \text{ and } a, \text{ constants.}$$

5. **Concluding remarks.** We have treated in this paper two of the simplest cases of uniform strain. When the strain is not uniform (as is, for example, the case throughout the interior of the earth) the problem is much more complicated. We content ourselves with pointing out here a certain inadequacy in the accepted treatment of what is, from the engineering standpoint, the most important case of non-uniform strain, namely, the uniform bending of a beam. In this accepted treatment $T(a) = \partial\{\rho(a)\phi\}/\partial\epsilon$ is assumed to be of the form

$$T(a) = \begin{pmatrix} 0 & 0 & 0 \\ 0 & 0 & 0 \\ 0 & 0 & P \end{pmatrix}$$

where P (instead of being constant as in the uniform tension of a cylinder) is a constant multiple of a^1. Under this assumption the basic equation of equilibrium div $(a)\{T(a)J^*\}$ yields $\partial^2 x/\partial a^3 \partial a^3 = 0$ so that each of the coordinates (x^1, x^2, x^3) must be a linear function of a^3. This implies that the fibers of the beam which are originally parallel to the axis of the beam remain straight lines after the strain (instead of being bent as in the accepted theory). In the classical theory J^* is set equal to E_3 before the differentiations involved in the symbol div (a) are performed and so the term involving the derivatives of the elements of J^*, to which div $(a)\{T(a)J^*\}$ reduces when div $(a)T(a) = 0$, is lost. The classical solution is furnished by the formulas[2]

[2] A. E. H. Love, *The mathematical theory of elasticity*, p. 127.

$$x = a + \frac{1}{2R}(c^2 + \sigma a^2 - \sigma b^2),\ y = b + \frac{\sigma}{R}ab,\ z = c - \frac{ac}{R}$$

where $P = -Ea/R$, R being a constant (σ = Poisson's ratio, E = Young's modulus). Thus

$$J^* = \begin{pmatrix} 1 + \sigma a/R & \sigma b/R & -c/R \\ -\sigma b/R & 1 + \sigma a/R & 0 \\ c/R & 0 & 1 - a/R \end{pmatrix}$$

so that the three matrices obtained by differentiating, in turn, the columns of J^* with respect to a are

$$\left(\frac{\partial J^*}{\partial a}\right)_1 = \begin{pmatrix} \sigma/R & 0 & 0 \\ 0 & -\sigma/R & 0 \\ 0 & 0 & 1/R \end{pmatrix};\quad \left(\frac{\partial J^*}{\partial a}\right)_2 = \begin{pmatrix} 0 & \sigma/R & 0 \\ \sigma/R & 0 & 0 \\ 0 & 0 & 0 \end{pmatrix};$$

$$\left(\frac{\partial J^*}{\partial a}\right)_3 = \begin{pmatrix} 0 & 0 & -1/R \\ 0 & 0 & 0 \\ -1/R & 0 & 0 \end{pmatrix}.$$

The traces of the products of each of these matrices by $T(a)$ should be zero; the first of these traces is $P/R = -Ea/R^2$ while the other two are zero.

THE JOHNS HOPKINS UNIVERSITY,
BALTIMORE, MD.

ON DYNAMIC STRUCTURAL STABILITY

BY

G. F. CARRIER

1. Introduction. Structures which are characterized by the fact that certain time-independent loadings (surface tractions) produce a state of instability are frequently subjected to dynamic loadings which also may produce an instability. We shall be concerned with such structures in this paper, considering only dynamic loadings which lead to a *stability problem* as contrasted with the *problem of forced vibrations*. We shall distinguish between these problems as clearly as possible. While treating this question it is convenient to adopt a new point of view in dealing with the *static* stability problem, and to modify its mathematical formulation accordingly. The results gained by this approach are readily shown to parallel rather closely those of W. Prager [4].

2. Formulation of the stability problem. Nearly all structures for which the question of structural stability must be raised are characterized by the presence of a typical dimension (shell thickness, rod diameter, and so on) which is small compared to the other typical dimensions of the system. For such structures, it is always possible to determine a system of surface tractions such that the displacements can be considered as small in the sense customary in the theory of elasticity.[1] These assumptions imply, essentially, that if a system of surface tractions T_i leads to a state of stress σ_{ij} and a displacement u_j, then λT_i should produce a stress $\lambda\sigma_{ij}$ and a deflection λu_j, for the range of λ to be considered. Any system T_i which obeys this restriction will be called *admissible*, and the symbol T_i^1 will be used to denote a particular admissible system of surface tractions. A simple example is furnished by a ring which assumes a noncircular shape under the influence of some surface traction T_i^0. A radial pressure q with distribution $q = \lambda/\rho$, where ρ is the radius of curvature in the pre-stressed configuration, satisfies our condition.[2]

We now define the dynamic stability problem. An elastic structure which is in equilibrium under the state of stress σ_{ij}^0 is subjected to surface tractions $\lambda T_i = T_i^1 A(t)$, where T_i^1 is defined in the foregoing and where $A(t)$ is some function of the time. A certain state of deformation λu_j^1 and a state of stress $\sigma_{ij}^0 + \lambda\sigma_{ij}^1$ result, which are both dependent on time. This state of stress and deformation may be found as the solution to a linear problem. For each value of λ, certain transient oscillations may exist, depending on the initial conditions. For certain ranges of λ and certain functions $A(t)$, these transients will not be stable;

[1] This is true even when the structure under consideration is already in a state of stress σ_{ij}^0 accompanied by large deformations. In this case, the quantities T_i, σ_{ij}, u_j are the *additional traction*, the *change in stress*, and the *deflection from the pre-stressed equilibrium configuration.*

[2] The static stability of such a configuration is discussed in [1].

that is, their amplitude will increase with time. When this occurs, the structure is said to be dynamically unstable. A discussion of this instability is the purpose of the present paper. Discussions of two particular cases of this phenomenon may be found in the literature [1, 2, 3, 5]. Since transients may always exist, further specification is required to make the definitions of u_j^1 and σ_{ij}^1 unique. This is accomplished by requiring that the state $\lambda\sigma_{ij}^1$, λu_j^1, which is produced by the application of $\lambda T_i^1 A(t)$, be that for which σ_{ij}^1, u_j^1 are independent of λ. The precise initial conditions to which this solution is related are unimportant and cannot, in general, be given in a less implicit form.

Let us now formulate the problem mathematically. We shall treat the quantities σ_{ij}^0, σ_{ij}^1, u_j^1 as known, in view of the fact that they are not strictly a part of the stability investigation. We shall adopt the conventional tensor notations according to which repeated subscripts imply summation and a comma followed by a subscript i denotes differentiation with regard to the rectangular Cartesian coordinate x_i. The configuration associated with the state σ_{ij}^0 is taken as the reference state and the Lagrangian manner of description will be adopted.

Denoting the states of stress and deformation (with transients admitted) by $\bar{\sigma}_{ij}$ and $\bar{u}_j$ respectively, we have for the condition of dynamic equilibrium

$$\bar{\sigma}_{ij,i} = \rho\partial^2\bar{u}_j/\partial t^2, \tag{1}$$

$$(\bar{\sigma}_{ij}\bar{x}_k - \bar{\sigma}_{ik}\bar{x}_j)_{,i} = \rho(\bar{x}_k\partial^2\bar{u}_j/\partial t^2 - \bar{x}_j\partial^2\bar{u}_k/\partial t^2). \tag{2}$$

When the force across an elementary surface dS is given by $d\bar{f}_i$, we have

$$d\bar{f}_j = \bar{\sigma}_{ij}n_i dS. \tag{3}$$

In the original state and in the state 1 which excludes transients, we have

$$\sigma_{ij,i}^0 = 0, \tag{4}$$

$$(\sigma_{ij}^0 x_k - \sigma_{ik}^0 x_j)_{,i} = 0, \tag{5}$$

$$df_j^0 = \sigma_{ij}^0 n_i dS, \tag{6}$$

$$(\sigma_{ij}^0 + \lambda\sigma_{ij}^1)_{,i} = \lambda\rho\partial^2 u_j^1/\partial t^2, \tag{7}$$

$$(\sigma_{ij}^0 + \lambda\sigma_{ij}^1)x_k^1 - (\sigma_{ik}^0 + \lambda\sigma_{ik}^1)x_{j,i}^1 = \rho(x_k^1\partial^2 u_j^1/\partial t^2 - x_j^1\partial^2 u_k^1/\partial t^2), \tag{8}$$

$$df_j^1 = (\sigma_{ij}^0 + \lambda\sigma_{ij}^1)n_i dS. \tag{9}$$

We also note the following fact. If $\bar{u}_j = u_j^1 + u_j$, then

$$\bar{x}_j = x_j + u_j^1 + u_j; \qquad \bar{x}_{j,l} = \delta_{jl} + u_{j,l}^1 + u_{j,l}. \tag{10}$$

We may also break up $\bar{\sigma}_{ij}$ into the parts

$$\bar{\sigma}_{ij} = \sigma_{ij}^0 + \lambda\sigma_{ij}^1 + \tau_{ij} + \tau_{ij}' + \tau_{ij}''. \tag{11}$$

Here, τ_{ij} and τ_{ij}' are, respectively, the symmetric and antisymmetric parts of the stress $\bar{\sigma}_{ij}$ which are associated with the strain $\epsilon_{ij} = (u_{i,j} + u_{j,i})/2$. The quantity τ_{ij}'' is associated with the rotation $\omega_{ij} = (u_{i,j} - u_{j,i})/2$.

We now follow the reasoning presented in detail in [4] and investigate the quantities τ, τ', τ''. It is clear that if a portion of the structure takes on a pure homogeneous strain $\epsilon_{ij} = u_{i,j} = u_{j,i}$, the change in stress $\tau + \tau'$ (since $\tau'' = 0$ in this case) arises in two ways. A symmetric change of stress τ_{ij} arises according to the generalized Hooke's law

$$\tau_{ij} = C_{ijkl}\epsilon_{kl}. \tag{12}$$

(Here C_{ijkl} is a fourth order tensor such that $C_{ijkl} = C_{jikl} = C_{ijlk}$.) As a rule, an anti-symmetric change of stress τ'_{ij} will then be required to preserve the moment equilibrium. To find the expression for τ'_{ij}, we combine Equations (1), (2), (7), (8), (10) and (11) and obtain[3]

$$\tau'_{ij} = -\tau'_{ji} = \{(\overset{0}{\sigma}_{ki} + \lambda\overset{1}{\sigma}_{ki})\epsilon_{jk} - (\overset{0}{\sigma}_{kj} + \lambda\overset{1}{\sigma}_{kj})\epsilon_{ik}\}/2. \tag{13}$$

In order to determine τ'' we note that under a pure rotation $\omega_{ij} = u_{i,j} = -u_{j,i}$, we have

$$d\bar{f}_i = (\delta_{ij} + \omega_{ij})df_j = df_i + \omega_{ij}\, df_j. \tag{14}$$

Using Equations (3), (9), (11) and (14) we obtain

$$\tau''_{ij} = \omega_{jk}(\overset{0}{\sigma}_{ik} + \lambda\overset{1}{\sigma}_{ik}). \tag{15}$$

A combination of Equations (1), (7) and (11) now yields

$$[\tau_{ij} + \tau'_{ij} + \tau''_{ij}]_{,i} = \rho\partial^2 u_j/\partial t^2. \tag{16}$$

With the boundary condition that $T_i^0 + \lambda T_i^1 = \bar{T}_i$, that is, $df_j = d\bar{f}_j$, on the surface, we have

$$[\tau_{ij} + \tau'_{ij} + \tau''_{ij}]n_i = 0. \tag{17}$$

Equations (12), (13), and (15) provide the expressions for the stresses τ, τ', τ'' in terms of the known stress systems $\overset{0}{\sigma}_{ij}$, $\overset{1}{\sigma}_{ij}$, and the deflection u_j. Thus equations (12), (13), (15), (16), and (17) imply a homogeneous differential equation in u_j and boundary conditions which will lead to the determination of the transients and their stability or lack of stability.

3. The variational formulation of the problem. In order to find the variational formulation of the problem, it is essential that we look more closely into the nature of the stress system $\overset{1}{\sigma}_{ij}$. We may decompose this system into three distinct components: σ_{ij}, σ'_{ij}, σ''_{ij}. These are completely analogous to the τ, τ', τ'' of the preceding discussion. Hence, when $\overset{1}{u}_{i,j} = \overset{1}{\epsilon}_{ij} + \overset{1}{\omega}_{ij}$, it follows that

$$\sigma_{ij} = C_{ijkl}\overset{1}{\epsilon}_{kl}, \qquad \sigma'_{ij} = (\overset{0}{\sigma}_{ik}\overset{1}{\epsilon}_{kj} - \overset{0}{\sigma}_{jk}\overset{1}{\epsilon}_{ki})/2, \qquad \sigma''_{ij} = -\overset{1}{\omega}_{kj}\overset{0}{\sigma}_{ik}.$$

We now note that, if $\sigma_{ij}/\overset{0}{\sigma}_{ij} \ll 1$, no appreciable additional loading is obtained by the application of T_i^1 and hence no instability will occur unless the struc-

[3] In order to be consistent with our assumption of small displacements from the initial state we must disregard the terms of higher order.

ture is already unstable in the state $\overset{0}{\sigma}_{ij}$. Accordingly, $\overset{1}{\sigma}_{ij}$ must have at least the same order of magnitude as $\overset{0}{\sigma}_{ij}$. On the other hand σ'_{ij} is necessarily small compared to unity. Hence, the stresses σ'_{ij}, σ''_{ij} are small as compared to $\overset{0}{\sigma}_{ij} \sim \overset{1}{\sigma}_{ij} \sim \sigma_{ij}$.[4] To be consistent we must neglect σ'_{ij}, σ''_{ij} as compared to σ_{ij}, since we superimpose displacements and make use generally of the linearity condition. However, σ_{ij} is symmetric and thus $\overset{1}{\sigma}_{ij}$ must be treated as a symmetric tensor. Using this fact we see that equations (16), (17) are the Euler equations and natural boundary conditions of the variational problem

$$(18) \qquad \delta \int [C_{pqrs}\, \epsilon_{pq}\, \epsilon_{rs} + \Sigma_{pq}(u_{r,p}\, u_{r,q} - \epsilon_{rp}\, \epsilon_{rq}) - \rho(\partial u_j/\partial t)^2]dv\, dt = 0,$$

where Σ_{pq} denotes $\overset{0}{\sigma}_{pq} + \lambda \overset{1}{\sigma}_{pq}$.

We thus formulated the dynamic stability problem in terms of a variational principle as well as in terms of a system of equations. Given the states of stress $\overset{0}{\sigma}_{ij}$ and $\overset{1}{\sigma}_{ij}$, we now have only to solve either equations (16), (17) or (18) for u_j.

A fairly obvious procedure for approximating such solutions is the following. We choose u_j in the form

$$(19) \qquad u_j = B_j(x_k)C(t)$$

where $B_j(x_k)$ is an (intelligently) guessed function of the coordinates. The spatial integration of (18) followed by the variation of the remaining integral leads to an ordinary differential equation in $C(t)$. In particular, if the time dependence of T_i^1 is sinusoidal, this equation for $C(t)$ will be a Mathieu equation. For better accuracy, of course, one uses a linear combination of such functions and obtains, by a Rayleigh-Ritz procedure, a system of simultaneous equations for the time dependent coefficients of the spatial functions.

4. **The static stability problem.** Before proceeding to demonstrate an application of the foregoing formulation of the dynamic stability problem, we shall consider the question of static stability. It is quite evident that each of the foregoing steps still applies when the loads and deflections are no longer time dependent. In that case all inertia terms vanish and our equations resemble closely those of W. Prager [4]. We note that in this time-independent form these equations imply the existence of eigenvalues λ and corresponding eigenfunctions u_j as contrasted with the "transients" arising in the dynamic problem. The following point of view must be adopted to solve a static stability problem. Let us suppose that the state of stress $\overset{0}{\sigma}_{ij}$ produced by the loading T_i^0 is given. The following questions may then arise.

(1) Is the given equilibrium configuration stable?

(2) How can one define and measure the degree of stability?

(3) What multiple μ of T_i^0 will lead to instability?

If the eigenvalue λ associated with the given state $\overset{0}{\sigma}_{ij}$ and the loading T_i^1 is zero, then the given configuration is unstable. If $\lambda > 0$, the configuration is

[4] The symbol $\sim$ should be read as "which is of order."

stable and λ might be used as a rough "safety factor" (we imply that the loading T^1_{ij} is so directed as to render the structure less stable). If T^0_i is not of the type T^1_i, question (3) must be answered by a sequence of problems. We guess a multiple μ_1 of T^0_i, compute the new σ^0_{ij} and configuration, and find λ_1 as we found λ. We proceed in this fashion, defining a sequence of $\mu_k = \mu(\lambda_k)$, until we find $\mu(0)$. This is the desired eigenvalue.

For purposes of comparison we state the point of view adopted by W. Prager. Here, we multiply the stress σ^0_{ij} by a constant λ *assuming no change in the configuration.* When we solve for the eigenvalue λ according to his equations [essentially our equations (16), (17), (18)], we may answer the three questions as follows. If $\lambda = 1.0$, the system is unstable and the answer is rigorous in view of the fact that the assumed configuration is the correct one only if $\lambda = 1$. If $\lambda > 1.0$ the system is stable and λ may be treated as a "safety factor". Question three is answered by defining a sequence of problems (as before) such that $\lambda_k \to 1.0$.

We note the following facts. If σ^0_{ij} is of the admissible type, the two points of view become identical and Prager's λ equals unity plus our λ. For each linear problem solved in obtaining the answer to question (3), our method produces a loading $T^0_i + T^1_i$ which defines an unstable state of stress and the corresponding configuration. This may or may not be of interest, of course.

5. **An example.** Let us apply the results of §2 to investigate the stability of a thin circular disc which is subjected to a uniform radial pressure.[5] Since this loading is of the type T^1_i, we may write

$$\lambda\sigma^1_{ij} = \lambda A(t) \begin{vmatrix} 1 & 0 & 0 \\ 0 & 1 & 0 \\ 0 & 0 & 0 \end{vmatrix},$$

to a sufficient degree of accuracy. In accordance with thin plate theory, we choose $u_3 = w(x, y)$ as the deflection of the neutral surface of the plate and write

$$\tau_{13} = \frac{E(h^2/4 - z^2)}{(1 - \nu^2)} \Delta w_x, \qquad \tau_{23} = \frac{E(h^2/4 - z^2)}{(1 - \nu^2)} \Delta w_y,$$

$$\omega_{13} = w_x, \qquad \omega_{23} = w_y, \qquad \epsilon_{13}/\omega_{23} \ll 1.$$

If we now let $j = 3$ in (16) and integrate over the thickness of the plate $(-h/2 \leqq z \leqq h/2)$, we obtain

$$\frac{Eh^2}{12(1 - \nu^2)} \Delta\Delta w + \lambda A(t)\Delta w = \rho\partial^2 w/\partial t^2. \tag{20}$$

Here Δ is the Laplacian operator.

When we consider the lowest mode of oscillation, w is readily seen to be of the

[5] We could as easily re-derive the known result for the column under dynamic end-thrust [2].

form $w = g(t)J_0(\alpha r)$ where αr_0 is the smallest root of the equation $J_0(p) = 0$ and where r_0 is the disc radius.[6] With abbreviated constants, (20) becomes

$$[k\alpha^2 - \lambda\alpha A(t)]g(t) = \rho\partial^2 g/\partial t^2. \tag{21}$$

Thus, the behavior of the function $g(t)$ defines the stability of the disc under the given loading. Note that if $A(t)$ is of the form $(a + b \cos \omega t)$, (20) is a Mathieu equation. A discussion of this equation which is relevant to the stability problem may be found in [2].

Bibliography

1. G. F. Carrier, *On the buckling of elastic rings*, Journal of Mathematics and Physics vol. 26 (1947) pp. 94–103.
2. S. Lubkin and J. J. Stoker, *Stability of columns and strings under periodically varying forces*, Quarterly of Applied Mathematics vol. 1 (1943) p. 215.
3. E. Mettler, *Über die Stabilität erzwungener Schwingungen elastischer Körper*. Ingenieur-Archiv vol. 13 (1942) pp. 97–103.
4. W. Prager, *The general variational principle of the theory of structural stability*, Quarterly of Applied Mathematics vol. 4 (1947) pp. 378–384.
5. Josef Taub, *Stossartige Knickbeanspruchung schlanker Stäbe im elastischen Bereich*, Luftfahrtforschung vol. 10 (1933) pp. 65–86.

Brown University,
Providence, R. I.

[6] This solution actually corresponds to the boundary condition that the turning at the edge is elastically resisted. We choose this boundary value problem in order that the solution appear in closed form. This seems justified since our purpose is merely to demonstrate an application of the theory.

STRESS-STRAIN RELATIONS FOR STRAIN HARDENING MATERIALS: DISCUSSION AND PROPOSED EXPERIMENTS[1]

BY

D. C. DRUCKER

A set of experiments is proposed which makes a sharp, easily measured distinction between the present groups of mathematical theories of plasticity for strain hardening materials. The results will not only determine the relative validity of these theories but will also provide the basis for a better mathematical description of the mechanical behavior of materials.

The basic experiments suggested for the strain hardening range are the rotation of the axes of principal stress without any change in the magnitude of the principal stresses. This can be accomplished by proper variation of tension, torsion, and internal pressure applied to a thin-walled tube.

It is believed that physical reasoning applied to the description of the experiments alone shows that the commonly accepted theories of plastic *deformation* can not be valid in general and that, to a much smaller degree, there is a limit to the range of validity of the theories of plastic *flow*. However, only the experimental results can provide the true answer for each material.

1. **Mathematical theories of plasticity.** To demonstrate that rotation without change in stress magnitude is fundamental, it is desirable to review the two main groups of mathematical theories of plasticity for strain hardening materials which have been formulated on the basis of limited experimental results. They have been termed [4, 6] theories of plastic *deformation* and theories of plastic *flow*. The first (deformation) states that the components of strain are determined completely by the existing components of stress and the second (flow) that the small increments in the components of strain depend upon the state of stress and the small increments in stress. Time effects which are important in many applications and for numerous materials are not taken into account in these mathematical formulations. This limitation is not removed in the following discussion.

Neither type of theory claims generality even for simple states of stress, and each can be applied in a straightforward manner only when the material is loaded continuously or when loading is followed by a single unloading. For example, in simple tension, loading followed by unloading produces a load-elongation or a stress-strain curve as shown in Fig. 1. Subsequent reloading retraces most of the straight line *CB* so that the new stress-strain relationship is not the same as for the original loading *AB*.

It is convenient and possibly necessary on physical grounds to consider the

[1] The conclusions presented in this paper were obtained in the course of research conducted under Contract N7onr-358 sponsored jointly by the Office of Naval Research and the Bureau of Ships.

components of elastic strain (recoverable) ϵ'_{ij}, and plastic strain (permanent), ϵ''_{ij} separately. The total strain ϵ_{ij} is

$$\epsilon_{ij} = \epsilon'_{ij} + \epsilon''_{ij}. \tag{1}$$

A11-3

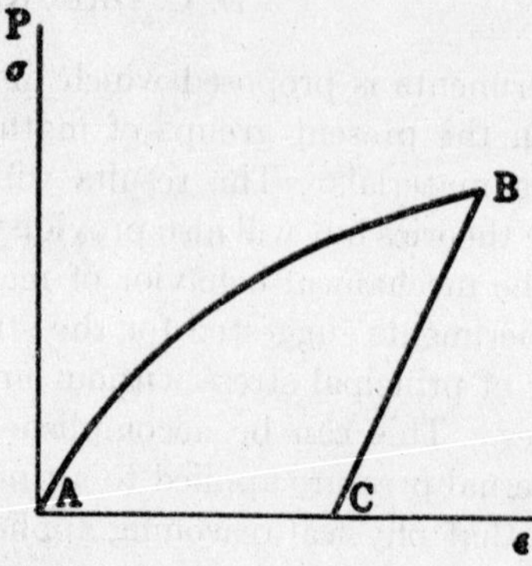

FIG. 1

Deformation theory. The often used relationship that octahedral shearing strain (either total or permanent) is a unique function of octahedral shearing stress is a special case of the general *deformation* theory.

The usual assumptions of *deformation* theory are that (a) strain is uniquely determined by stress, (b) the material is isotropic so that the principal axes of stress coincide with the principal axes of permanent strain,[2] (c) the ratios of the principal strains will be constant if the ratios of the principal stresses are kept constant during the test, and (d) change in volume is a purely elastic phenomenon. Prager [8] has shown that, in accordance with these assumptions, the most general mathematical form of expression for the permanent strain when loading (only) takes place is

$$\epsilon''_{ij} = F(J_2, J_3^2)[P(J_2, J_3^2)s_{ij} + Q(J_2, J_3^2)J_3 t_{ij}] \tag{2}$$

where, with the usual summation convention for repeated subscripts:

$J_2 = s_{ij}s_{ji}/2$ is proportional to the square of the octahedral shearing stress.

$J_3 = s_{ij}s_{jk}s_{ki}/3$ is the third order invariant of the stress deviation; J_2 is the second.

$s_{ij} = \sigma_{ij} - s\delta_{ij}$ is the stress deviation.

$s = \sigma_{ii}/3$ is the average principal stress.

σ_{ij} is the stress.

$t_{ij} = s_{ik}s_{kj} - 2J_2\delta_{ij}/3$ is the deviation of the square of the stress tensor.

Flow theory. The *flow* theory is similar in form with the important distinction that differential changes are considered and ϵ''_{ij} is not determined by the final

[2] The permanent strains considered are small compared with unity but may be large compared with the elastic strains.

values of stress but depends upon the cumulative effect, that is, the path of loading.

If the assumptions [2] are made that (a) the previous strain history is not important, (b) the changes in strain are determined uniquely by the state of stress and changes in stress, and (c) the principal axes of the permanent strain increments will coincide with the principal axes of the existing stress system, it follows that

(3) $$d\epsilon''_{ij} = [p(J_2, J_3^2)s_{ij} + q(J_2, J_3^2)J_3 t_{ij}]\, dJ(J_2, J_3^2)$$

where the invariants of the stress deviation, J_2 and J_3, appear and the change in plastic strain depends upon the change in J_2 (or, equivalently, upon the change in the octahedral shearing stress) and the change in J_3^2.

Comparison. Both types of theories can and should be made to predict that during unloading there is no change in the permanent strain and that during loading or unloading the elastic strains are functions of the stresses, for example, Hooke's law for an isotropic material

(4) $$E\epsilon'_{ij} = (1 + \nu)s_{ij} + (1 - 2\nu)s\delta_{ij}$$

where E is Young's modulus and ν is Poisson's ratio.

Basically, the particular or the general formulations of the two groups of theories conflict mainly in their directional properties. *Deformation* theory predicts coincidence of the axes of stress and strain, *flow* theory does not. They must, therefore, lead to different results in most cases including all except the simplest practical problems involving structural members or machine elements. However, they can be made to give the same answer when the directions of the principal stresses are kept constant and the ratios of the stress magnitudes remain fixed as loading proceeds, that is, if

(5) $$\sigma_{ij} = K\sigma^0_{ij},$$

where σ^0_{ij} is any given state of stress and K is a scalar variable which starts at zero and increases continuously.

Unfortunately, this special agreement between the two theories exists in almost all of the tests that have been performed to determine the plastic behavior of materials.[3] Plots of octahedral shear stress against octahedral shear strain and also maximum shear stress against maximum shear strain correlate very well under these conditions of testing. However, there is little reason to suspect nor does it seem at all likely that a truly general relationship actually exists between these or similar quantities.

2. Rotation of the axes of principal stress. The basic experiment to determine the validity of the theories is to change the directions of the axes of principal stress in the strain-hardening range without altering the magnitude of the stresses. Such a change has been termed a neutral change in stress [2]. It constitutes

[3] The tests of Taylor and Quinney [9] which will be discussed later are actually not exactly of this type although they may seem to be on first inspection.

neither loading nor unloading in the general sense, because all the stress invariants (for example, the octahedral shear stress) remain the same.

The predictions of the two theories are clearly distinct and entirely different. If the permanent strain depends upon the state of stress alone (isotropy is necessarily implied) the principal plastic strains must coincide in direction with the principal stresses at all times. *Deformation* theory therefore predicts that the permanent strain system will rotate with the principal stresses. On the other hand, the *flow* theory predicts no change in any component of permanent strain, $d\epsilon''_{ij} = 0$, because both the octahedral shear stress and J_3 are constant, $dJ = 0$, equation (3).

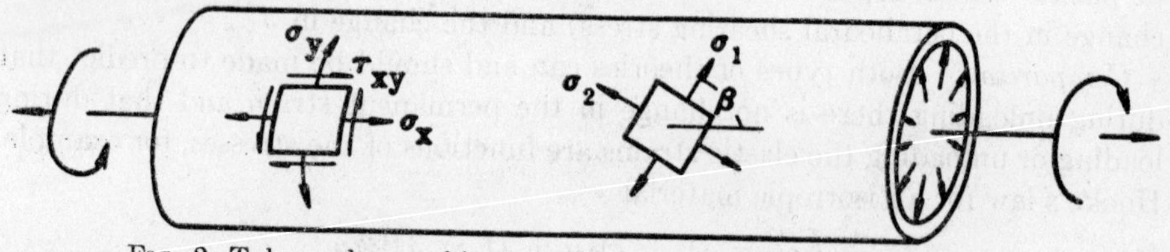

FIG. 2. Tube under combined tension, torsion, and internal pressure

For appreciable initial plastic strains and subsequent rotation about the intermediate, usually also about the smallest principal stress direction, the difference between these two predictions is so large that little precision of measurement is required to determine which theory is closer to the physical facts.

It is of interest to note that previous [7][4] suggestions for experimental work to distinguish between the theories actually depended fundamentally on the difference in prediction about the effect of rotation, but often were somewhat obscured by other changes taking place simultaneously.

Physical reasoning suggests strongly that, for small rotations of the principal stress directions, the strain hardening which has previously taken place will preclude appreciable additional slip, and therefore the *deformation* type of theory cannot be correct. However, a 90° rotation will cause a high shear stress to act on a set of planes that are still weak and further slip may occur, that is, strain hardening is not isotropic and the *flow* theory is not complete. The preceding reasons are too close to hypotheses and the experiments must be carried out to determine what actually takes place.

3. **Proposed experiments.** An experiment involving plastic strain cannot be interpreted easily unless the stress and strain are reasonably homogeneous. A thin tube under combined tension, torsion, and internal pressure is the only test specimen so far devised which permits a variety of approximately plane states of stress σ_x, σ_y, τ_{xy} (Fig. 2). However, compressive stresses are severely limited by the possibility of buckling, especially in the plastic range.

The most striking and probably most significant test is the rotation, in the plastic range, of a state of uniaxial tension as shown in Fig. 3. The loading

[4] Further references are given in this paper.

schedule is obtained by gradually changing β in the expressions for the stress components

(6) $$\sigma_x = \frac{\sigma}{2} + \frac{\sigma}{2}\cos 2\beta, \quad \sigma_y = \frac{\sigma}{2} - \frac{\sigma}{2}\cos 2\beta, \quad \tau_{xy} = \frac{\sigma}{2}\sin 2\beta.$$

Starting with $\beta = 0$, that is, σ in the x direction, requires simply a gradual application of tensile force only to the tube until the tensile stress σ is reached. Then

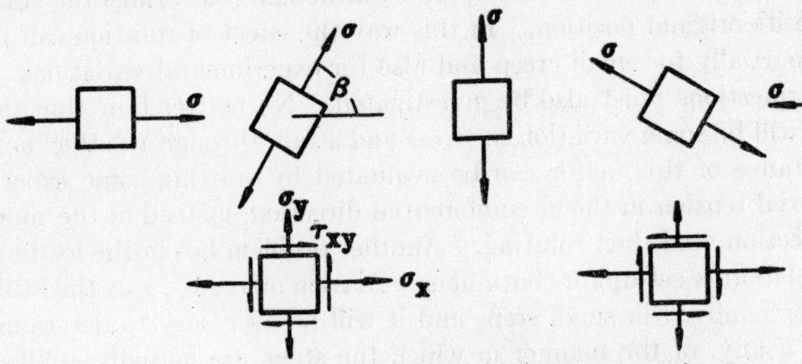

Fig. 3. Rotation of a uniaxial tension

β is increased slowly which means decreasing the axial tension and adding internal pressure to give σ_y and torque to give τ_{xy}. The torque increases to a maximum at $\beta = 45°$ and decreases to zero at $\beta = 90°$, reverses in direction increasing in magnitude to $\beta = 135°$ and decreasing to zero again at $\beta = 180°$. The internal pressure can be made to produce circumferential stress alone [1] or the more usual boiler type of axial and circumferential stress. Circumferential alone is preferable in that $\beta = 90°$ can then be obtained solely by internal pressure; otherwise an axial compressive force must be added to eliminate the axial tension.

Strain indicating devices will measure the changes in both the elastic and permanent strains and the two should be separated if an accurate theory is to be developed. However, when the permanent strain is, say, ten times the elastic, the distinction between the flow and deformation theories is so large that even this simple correction is not required.

Any other state of plane stress is obtained by simply adding a two-dimensional all around tension in the surface of the tube to the σ system. This requires added axial force and internal pressure only to give $\sigma_x = \sigma_y = \sigma'$, $\tau_{xy} = 0$. The principal stresses in the surface are $\sigma + \sigma'$ in the direction of σ, that is, at β to the x axis, and σ' at right angles. Note that the plane "hydrostatic" case is not the same as true, three-dimensional, all around tension or compression which produces no change in permanent deformation. Here the third principal stress is zero and the added σ' means an increase of $\sigma'/2$ in the maximum shear stress on the planes at 45° to the surface.

Despite the fact that the possible states of stress [9] cover the entire range $-1 \leqq \mu \leqq +1$ where $\mu = 2(\sigma_2 - \sigma_3)/(\sigma_1 - \sigma_3) - 1$, they are not really complete from the physical point of view because rotation is always about the direction of the smallest principal stress. It would be desirable, for example, to rotate a state of pure shear in the surface, but buckling difficulties prohibit this for thin tubes and lack of homogeneity of stress rules out thick tubes.

The first series of tests for each material should be rotation of uniaxial tension at, say, three values of σ in the strain-hardening range each starting with an initially unstrained isotropic tube (see Taylor and Quinney test for isotropy [9]). It is recommended that 360° rotation be studied although 180° brings the state of stress back to its original position. In this way the effect of rotation can be used to correct partially for small creep and also for experimental variations.

Several other questions must also be investigated. No matter how thin the tube wall, there will be some variation in stress and strain through the thickness [5]. The importance of this factor can be evaluated by starting some experiments with uniaxial tension in the circumferential direction, instead of the more natural axial direction, and then rotating. Another problem lies in the loading. Without a very elaborate set-up for continuous variation of σ_x, σ_y, τ_{xy}, the loads will be applied or removed in small steps and it will be necessary to determine the importance, if any, of the manner in which the steps are actually applied. This can be accomplished by exaggerating the steps in a few tests and also by running some tests in which σ is made to increase slightly at all times and others in which σ continuously decreases a slight amount.

The next series of tests could be run at $\sigma' = \sigma$ or $\sigma_1 = 2\sigma_2$ and possibly a third series at σ_1 just slightly larger than σ_2 which will further check the importance of creep and the precision of the experiments. The rotation should have little effect in the third series because of isotropy in the surface.

It should be emphasized again that the determination of the relative validity of the theories of plastic deformation and of plastic flow requires a very few experiments only, but that a good physical basis for a more complete theory needs both a great variety and repetition of test results. Only the test results can indicate how many series and repetitions are necessary. Also further studies involving loading after or during rotation, to provide a basis for a more general law of plasticity, cannot be outlined until the rotation information is carefully analyzed.

However, it is interesting and important to note that if the prediction of the flow theory, that is, $d\epsilon''_{ij} = 0$, is reasonably close for rotations up to 45° then the work of Taylor and Quinney [9] may be used to supply much of the additional information. In their experiments, tubes were stretched plastically in axial tension, some or all of the tensile force was released and torque applied. They obtained a consistent interpretation of their results by considering only the added strains produced by the loading and ignoring the original tensile strain, that is, by adopting in a sense a *flow* theory point of view. Therefore, their conclusions do not support the ordinary *deformation* theory as is so often assumed, but instead, what might be termed a flow-modified deformation theory. Further evidence in favor of such a modified deformation theory is given in detail by Hohenemser and Prager [3].

4. **Conclusion.** The rotation experiments will clearly distinguish between and determine the validity of the theories of the plastic behavior of strain-hardening materials. In addition, the results should prove extremely useful in providing a start for a more general theory of plasticity, should one be required, because any change in the state of stress can be considered as a change in the magnitude of the principal stresses plus a rotation of their directions.

Bibliography

1. E. A. Davis, *Yielding and fracture of medium-carbon steel under combined stress*, Journal of Applied Mechanics, Transactions of the American Society of Mechanical Engineers vol. 67 (1945) p. A-15.

2. G. H. Handelman, C. C. Lin, and W. Prager, *On the mechanical behavior of metals in the strain-hardening range*, Quarterly of Applied Mathematics vol. 4 (1947) pp. 397–407.

3. K. Hohenemser and W. Prager, *Beitrag zur Mechanik des bildsamen Verhaltens von Flusstahl*, Zeitschrift für Angewandte Mathematik und Mechanik vol. 12 (1932) pp. 1–14. (Available as RTP Translation No. 2468, Durand Reprinting Committee.)

4. A. A. Ilyushin, *Relation between the theory of Saint-Venant-Lévy-Mises and the theory of small elastic-plastic deformations*, Prikladnaia Matematika i Mekhanika vol. 9 (1945) pp. 207–218.

5. W. R. Osgood, *Combined stress tests on 24 S-T aluminum alloy tubes*, Journal of Applied Mechanics, Transactions of the American Society of Mechanical Engineers vol. 69 (1947) pp. A-147–153.

6. W. Prager, *An introduction to the mathematical theory of plasticity*, Journal of Applied Physics vol. 18 (1947) pp. 375–383.

7. ———, *Exploring stress-strain relations of isotropic plastic solids*, Journal of Applied Physics vol. 15 (1944) pp. 65–71.

8. ———, *Strain hardening under combined stresses*, Journal of Applied Physics vol. 16 (1945) pp. 837–840.

9. G. I. Taylor and H. Quinney, *The plastic distortion of metals*, Philos. Trans. Roy. Soc. London Ser. A. vol. 230 (1931) pp. 323–362.

BROWN UNIVERSITY,
PROVIDENCE, R. I.

THE EDGE EFFECT IN BENDING AND BUCKLING WITH LARGE DEFLECTIONS

BY

K. O. FRIEDRICHS

While perhaps the majority of the talks given at this Symposium were concerned with non-linear problems in fluid dynamics, I should like to discuss a class of problems in the theory of non-linear elasticity concerned with bending and buckling of plates and shells subjected to loads which are so large that the linear theory of elasticity is no longer applicable. On the other hand, the deflections are not so large that the theory developed by Murnaghan must be applied because the assumption that the strains remain small will still be maintained.

This field of bending and buckling of plates and shells attracted considerable attention up to four or five years ago; the purpose of my talk is not to present new material but to try to revive interest in this field. There is a certain aspect of these problems which I think is particularly fascinating, namely the behavior of such bent and buckled plates near the edge, and I shall concentrate on this aspect. Strange phenomena at the boundary of bent plates and shells can always be expected if the thickness of the plate or shell is very small. It should be noticed that this is already the case when the theory of linear elasticity is applicable. There are, however, certain peculiarities of this "edge effect" which apparently occur only in non-linear problems. One can definitely say they are due to the non-linearity. Nevertheless, I should first like to make a few remarks concerning the edge effect in linear problems.

The "edge effect" was first discovered by H. Reissner [6] in connection with the bending of shells on the basis of a linear theory. I prefer to discuss it in connection with a somewhat simpler problem.

Consider a flat plate of thickness h which is subjected to a lateral load per unit area hf and also a tensile constant stress σ in the plane of the plate. Its lateral deflection w is characterized by a differential equation of the form

$$(1) \qquad \gamma^2 h^2 \nabla^4 w - \sigma \nabla^2 w = f, \qquad \gamma^{-2} = 12(1 - \nu^2), \qquad E = 1,$$

and appropriate boundary conditions. Suppose the plate is very thin, that is, that h is very small. Then one is tempted to treat the problem approximately by setting $h = 0$. Mathematically speaking that means one considers a set of problems depending upon the parameter h and asks whether the solutions approach a solution of the problem when h is set equal to zero. An obvious difficulty arises. The original differential equation is of the fourth order and accordingly two boundary conditions should be imposed at the edge. The limiting differential equation is only of the second order and consequently one condition only can be imposed for the solution of the limiting differential equation. Assuming that it is true that the solutions for $h \neq 0$ approach the solution for $h = 0$ one would like to be able to characterize this limit function intrinsically as a solution of the limiting second order differential equation satisfying one proper boundary condition. The question then arises, which is this boundary condition or, in other words, which boundary condition got lost? There is a general rule which gives the answer to this question which has been completely established for ordinary differential equations by W. Wasow [8]. In the present case the rule shows that the boundary condition of the highest order of differentiability is lost. Suppose originally the plate was clamped so that the deflection w and the slope $\partial w/\partial n$ were prescribed to be zero at the edge. Then the latter condition will be lost. The limiting function will not have a vanishing slope at the edge. It is then clear that the first derivatives of the solution for $h \neq 0$ cannot converge uniformly up to and including the edge to the first derivatives of the limiting solution since these derivatives vanish for the solutions with $h \neq 0$, but not for the limit solution with $h = 0$. As a matter of fact, the convergence is non-uniform at the edge and for small values of the thickness h there is a very narrow strip along the edge at which the slope of the deflection quickly changes over from the value zero prescribed at the edge to a value not differing much from that which the limiting solution assumes at the edge. This narrow strip corresponds to the boundary layer in the flow of a viscous fluid along an obstacle and it is frequently also called a boundary layer.

For many purposes it is important to know just how this rapid transition across the narrow boundary layer takes place. This question can be answered in a manner similar to that employed in Prandtl's boundary layer theory. Instead of the arc length s and the normal n, new independent variables hs and hn are introduced so that as $h \to 0$ the boundary layer is magnified and in the limit extends to infinity. The transformation is so chosen that in the limit the order of the differential equation does not reduce. If now w is considered a function of hn and hs, a completely new limiting procedure results in which the convergence is uniform. The limiting differential equation problem is simpler since the domain is the half-plane. The only question this time is: what are the appropriate boundary conditions at infinity by which this limiting solution can be characterized? The answer is, roughly speaking, that the value of the quantity dw/dn at infinity is the same as that which the solution of the interior limit process, described before, assumes at the edge.

This is an outline of the theory of the edge effect problems of bending in which linear elasticity is still applicable. The question arises whether the situation is not the same in non-linear problems. I want to show that in certain non-linear problems the situation is the same and in certain other problems it is quite different.

Let us consider again a plate subjected to a lateral load and a tensile load in the plane of the plate applied at the edge. But now we no longer assume that the resulting deflection w is so small that the theory of linear elasticity is applicable but still we assume that the strains and slopes remain small. For that case von Kármán [4] has formulated the basic equations

(2) $\gamma^2 h^2 \nabla^4 w - \varphi_{yy} w_{xx} + 2\varphi_{xy} w_{xy} - \varphi_{xx} w_{yy} = f, \qquad \nabla^4 \varphi + w_{xx} w_{yy} - w_{xy}^2 = 0$

in which φ is Airy's stress function and hf is the lateral load per unit area. For simplicity let us assume that the plate is circular in shape and that the deflection has cylindrical symmetry depending only on the distance r from the center. It is then convenient to introduce instead of w the quantity

(3) $$q = -r^{-1} w_r$$

proportional to the slope of the plate and further the radial stress component σ counted positive when it is a tension. The differential equations then can be reduced to the following form

(4) $$\gamma^2 h^2 r^{-3}(r^3 q_r)_r - \sigma q = -f/2,$$

(5) $$r^{-3}(r^3 \sigma_r)_r + q^2/2 = 0.$$

Two boundary conditions must now be imposed, one for q and one for σ. We require that the plate is free at the edge in the sense that the bending moment vanishes at the edge:

(6) $$rq_r + (1 + \nu)q = 0, \qquad r = R.$$

Secondly we prescribe the value of the radial stress σ at the edge

(7) $$\sigma = \bar{\sigma}, \qquad r = R.$$

We first consider the simple case that the edge stress $\bar{\sigma}$ is prescribed to be zero while the load is different from zero, $f \neq 0$, and we want to find approximately the solution of the problem in case the thickness h is very small. It is then natural simply to set $h = 0$ in equation (4). From the resulting equation one can immediately express q in terms of σ and insert in the equation (5). The resulting equation

(8) $$r^{-3}(r^2 \sigma_r)_r + f^2/8\sigma^2 = 0$$

can easily be solved numerically. This problem is analogous to a problem treated by Hencky. The question arises whether the solutions for small h actually approach the solution of the present problem for $h = 0$. There is little doubt that this is the case. Also there is little doubt that the lost boundary condition is condition (6) concerning the bending moment. Thus, for small values of h, there will be a narrow layer in which the bending moment changes quickly from the value zero at the edge to approximately the value which the solution of the limiting problem assumes at the edge, a value which, by the way, is infinite.

There is another curious phenomenon met in this case due to the fact that the second term in equation (8) becomes infinite at the edge. The rate of change of the stress σ also becomes infinite at the edge. Thus also the pressure undergoes a very quick change near the edge by (7). Experimentally this behavior could not be distinguished from a boundary layer but mathematically speaking it is a completely different affair.

Now let us consider the case that the load hf vanishes and that the stress $\sigma = \bar{\sigma} = -\bar{p}$ at the edge is a compression. If the pressure $\bar{p}$ exceeds a certain value the plate buckles out and, if the thickness h of the plate is very small, this happens already for a very small value of the edge pressure $\bar{p}$. Again one would like to try to describe the buckled state of the plate approximately by simply setting $h = 0$ in (4). If one does, the result is the equation

$$\sigma q = 0 \tag{9}$$

with the solution $q = 0$, whence from (5) $\sigma =$ const. follows. It is again natural to assume that the condition (6) concerning the bending moment is lost and that condition (7) remains valid. Condition (6), by the way, is satisfied automatically by the limiting solution. If now condition (7) were retained the limit solution would simply be that the constant stress σ equals that which was prescribed at the edge. However, that is just not correct. The condition (7) concerning the edge stress is also lost in the limit and the correct limit solution does not satisfy it. The correct limiting solution is, as a matter of fact, the constant

$$\sigma = -.47\bar{p}. \tag{10}$$

Thus the limit process in this case is irregular and differs completely from what has been established to be generally true for corresponding limiting processes for linear equations. In the normal limiting process the direct limit function could be determined intrinsically and provide information which was needed to determine the limit function of the process of stretching the boundary layer. Here, however, the situation is the reverse. It can be shown that the stretching of the boundary layer leads to a problem through which the boundary layer limiting solution is determined intrinsically and which provides information needed to determine the direct limit solution. This is the way the number $-.47$ was obtained. This reversal of the determinacy situation appears to be due to the non-linearity.

This peculiar situation makes it quite imperative to establish the statements made by a rigorous mathematical proof, which has been supplied by Stoker and myself years ago [3], but in addition it makes one suspicious about similar situations and one feels that even in cases in which one expects a regular limiting behavior a rigorous proof is called for. Thus for the problem mentioned before for the bending of a plate such a proof is under investigation.

To determine approximately the solutions of these problems is not difficult once one knows their qualitative behavior. One may thus, for example, use a Rayleigh-Ritz procedure as has been proposed by von Kármán. However, unless one knows this asymptotic behavior beforehand such approximation methods might give quite erroneous results.

A very famous problem for which an approximate solution can be given once the correct asymptotic behavior is properly conjectured is the following: a rectangular plate, simply supported along two sides, is subjected to a constant perpendicular displacement in the plane of the plate on the two other sides and then buckles out. To determine the deflection and the stress displacement for a very thin plate von Kármán [4] has assumed that boundary layers occur along the simply supported sides outside of which the stresses are negligible. On the basis of that conjecture von Kármán gave easily an approximate expression for the approximate stress. The width of the boundary is what he calls the effective width.

Another problem in which one may expect that a boundary layer approach will give the solution is that of the buckling of a sphere which is subjected to a uniform compressive load from the outside. If this load exceeds a certain value the sphere will buckle. As von Kármán and Tsien have shown [5], even for loads at which the unbuckled sphere is still stable, there exist states of equilibrium possessing a lower potential energy than the unbuckled state and they have advanced the hypothesis that the shell, before reaching instability, will jump over into one of these states of equilibrium. In these states of equilibrium the linear theory of elasticity is no longer applicable and it is rather difficult to determine them. Experimental evidence indicates that in the buckled state the shell develops one or perhaps several small dimples. If the thickness of the shell is very small, one is again tempted to try a limit process by letting $h \to 0$. One way of doing this would be to assume that as the thickness approaches zero the dimple contracts to a point and, accordingly, to magnify the neighborhood of this point by stretching it in proper relation to the thickness as it approaches zero. Thus the neighborhood of this point is stretched out into the full tangent plane at this point. In this way the mathematical formulation of the problem becomes much simpler and it then is amenable to a Rayleigh-Ritz procedure. Making a natural assumption about the shape of the dimple one obtains a numerical value for the buckling load which agrees nicely with the experimental value. However, it turns out that one can make the potential energy still smaller by making various unnatural assumptions about the shape of the dimple and it is still an open question whether or not the problem can be attacked in this way [2]. I should mention, though, that Tsien [7] has emphasized the fact that one should distinguish between two problems, one in which the external pressure is prescribed and another in which the change of volume is prescribed. The latter problem corresponds even better than the first one to the actual experimental situation and it seems that Tsien's treatment of the problem with prescribed volume change, in which he also makes use of the simplification by stretching, leads to a correct approximate result.

Finally I should like to mention a problem in which the boundary effect is just due to the non-linearity. This interesting phenomenon was observed by Stoker and Bromberg [1]. The problem is concerned with the deformations of a section of a sphere in which bending is neglected from the outset. Such a shell is called a sheet. The deformation of a sheet may be described by a radial displacement w and a tangential displacement u. Instead of the latter, one may conveniently introduce the stress component σ along a meridian curve. The differential equations of the problem are

$$R^{-1}(\sigma \sin \theta\, w_\theta)_\theta + \tan \theta(\sigma \sin \theta)_\theta + \sigma \sin \theta = -Rf, \tag{11}$$

$$(\tan \theta \sec \theta(\sigma \sin\theta)_\theta)_\theta - (1 + \nu \tan^2 \theta)\sigma + R^{-1}\{w \tan^2 \theta + w_\theta \tan \theta + (2R)^{-1} w_\theta^2\}. \tag{12}$$

As boundary conditions we require that the radial and circumferential deflection displacements w and u vanish at the edge. These conditions can be seen to be equivalent to

$$w = 0, \tag{13}$$

$$(\sigma \sin\theta)_\theta - \nu\sigma \cos\theta = 0. \tag{14}$$

If now the applied load hf is zero the solution of the problem is easily seen to be $w = 0$, $\sigma = 0$. To obtain a representation of the solution for small loads hf one may divide the quantities w and σ by f and then let $f \to 0$. One observes that if w/f and σ/f are introduced in equations (11) and (12), the first term in the first equation and the last term in the last equation obtain the factor f. Thus these two terms drop out in the limit. Incidentally, the resulting limit equations are just those of the so-called membrane theory of shells. Evidently the differential equation (11) which for $f \neq 0$ is of the second order is now an equation of the first order and accordingly one must expect that the boundary condition for w is not satisfied by the limit solution. This is indeed the case for small values of f. There is a narrow layer along the edge across which the ratio w/f rapidly changes from the value zero at the edge to approximately the value the limit function assumes at the edge. It is interesting to note that the data of concrete actual problems just fall into the range for which the boundary layer is very narrow. We observe that the first term in equation (11) is quadratic in the two unknown quantities σ and w and thus it is seen that the loss of a boundary condition and the occurrence of a boundary layer is due to the non-linearity of the problem.

Bibliography

1. E. Bromberg and J. J. Stoker, *Non-linear theory of curved elastic sheets*, Quarterly of Applied Mathematics vol. 3 (1945) pp. 246–265.

2. K. O. Friedrichs, *On the minimum buckling load for spherical shells*, Applied Mechanics, Theodore von Kármán Anniversary Volume, 1941, pp. 258–272.

3. K. O. Friedrichs and J. J. Stoker, *Buckling of the circular plate beyond the critical thrust*, Journal of Applied Mechanics vol. 9 (1942) pp. A7–A14. *The nonlinear boundary value problem of the buckled plate*, Amer. J. Math. vol. 63 (1941) pp. 839–888.

4. Th. von Kármán, *The engineer grapples with non-linear problems*, Bull. Amer. Math. Soc. vol. 46 (1940) pp. 615–683.

5. Th. von Kármán and H. S. Tsien, *The buckling of spherical shells by external pressure*, Journal of the Aeronautical Sciences vol. 7 (1939) pp. 43–50.

6. H. Reissner, *Spannungen in Kugelschalen (Kuppeln)*, H. Mueller-Breslau Festschrift, 1912, pp. 181–193.

7. H. S. Tsien, *A theory of buckling of thin shells*, Journal of Aeronautical Sciences (1942) pp. 373–384.

8. W. Wasow, *On the asymptotic solution of boundary value problems for ordinary differential equations containing a parameter*, Journal of Mathematics and Physics vol. 23 (1944) pp. 173–183.

New York University,
New York, N. Y.

NUMERICAL METHODS IN THE SOLUTION OF PROBLEMS OF NON-LINEAR ELASTICITY

BY

WILFRED KAPLAN

1. Form of the problem. If the stress-strain relations of a material are assumed to be non-linear and the differential operators are replaced by difference operators, then the equilibrium equations for the body considered become a system of n simultaneous non-linear equations in n unknowns:

$$f_i(x_1, \cdots, x_n) = 0 \qquad (i = 1, \cdots, n). \tag{1}$$

In the linear case, these equations represent the conditions for the minimum of a function $g(x_1, \cdots, x_n)$, the f_i being the components of the gradient of g.[1]

2. Methods of relaxation and gradients as special cases of a general procedure. A general method of obtaining the solution of (1) is to consider the corresponding ordinary differential equations

$$\frac{dx_i}{dt} = f_i(x_1, \cdots, x_n) \qquad (i = 1, \cdots, n). \tag{2}$$

These "dynamic" equations have an equilibrium point determined by (1). In the linear case all solutions are asymptotic to this point; for sufficiently small non-linearity this must continue to be the case, and that assumption will be made. Thus the determination of one solution of (2) will be sufficient.

The method of gradients amounts precisely to that. Because of the stability of the equilibrium point, one can further be satisfied with an approximate solution, for example, a broken-line solution. If a minimizing function g is available, then, in determining the length of each step, one can use the function g itself, and proceed in the chosen direction as long as g continues to decrease.[2]

The method of relaxation can also be interpreted as a method for approximate solution of (2). In this case one starts at an arbitrary initial point $(x_1^0, \cdots, x_n^0)$ and determines which, positively or negatively directed, coordinate axis, say x_1, is closest in direction to the vector f_i at the point, and proceeds some distance along that coordinate direction, perhaps until the "residual" f_1 equals 0.

In either of these two methods, the evaluation of the non-linear part of the f_i is lengthy, and, relying basically on a Picard successive approximation procedure, one usually reevaluates these only after several steps have been completed.

3. A compromise method. The writer has been experimenting with a method of approximate integration of (2) which permits more flexibility between the two extremes of the relaxation method and the method of gradients. The vital

[1] Cf. R. Courant, *Variational methods for the solution of problems of equilibrium and vibrations*, Bull. Amer. Math. Soc. vol. 49 (1943) pp. 1–23.

[2] Cf. H. B. Curry, *The method of steepest descent for non-linear minimization problems* Quarterly of Applied Mathematics vol. 2 (1944) pp. 258–261.

question is the accuracy with which the vector f_i is approximated at each step. The cruder the approximation, the more rapid the individual steps but (in general) the slower the convergence to equilibrium.

The basis of the new method is, first of all, a cubic lattice of mesh h in the $x_1x_2 \cdots x_n$-space, as in relaxation. An arbitrary initial point $(x_1^0, \cdots, x_n^0)$ is chosen and the vector $[f_1, \cdots, f_n]$ is evaluated at the point. The one of the vectors $[e_1, \cdots, e_n]$ nearest in direction to $[f_1, \cdots, f_n]$, where $e_i = \pm 1$ or 0, is then determined, and $(x_1^0, \cdots, x_n^0)$ is replaced by the lattice point $(x_1^0 + e_1h, \cdots, x_n^0 + e_nh)$ as next approximation. In two dimensions, this would amount to choosing the lattice direction within $22\frac{1}{2}°$ of the given vector, whereas the relaxation procedure would use 45°. The accuracy of the new procedure can be increased by allowing e_i to vary over a larger range, while decreasing h; for example, let $e_i = 0, \pm 1, \pm 2$, with h reduced one-half. In the case of linear equations, the process of changing many x's simultaneously can be greatly accelerated, for example, by use of punched cards. In the non-linear case, a similar remark applies if the usual procedure of correcting the non-linear terms only at intervals is used.

4. **An example.** Suppose one is solving the problem $\nabla^4 F = 0, F = F(x, y)$, on the square $0 \leqq x \leqq 3, 0 \leqq y \leqq 3$, with boundary conditions as shown in Table 1. The problem is then to be solved by difference equations on the lattice with mesh 1 unit. There are thus four unknown values: $x_1 = F(1, 2)$, $x_2 = F(2, 2)$, $x_3 = F(1, 1)$, $x_4 = F(2, 1)$. These satisfy the linear equations $a_{i\alpha}x_\alpha = c_i$, where, in matrix notation,

$$\| a_{ij} \| = \left\| \begin{matrix} 22 & -8 & -8 & 2 \\ -8 & 22 & 2 & -8 \\ -8 & 2 & 22 & -8 \\ 2 & -8 & -8 & 22 \end{matrix} \right\|, \qquad \| c_i \| = \left\| \begin{matrix} 2 \\ 31 \\ 15 \\ 14 \end{matrix} \right\|.$$

The exact solution is $x_1 = 1.575$, $x_2 = 2.775$, $x_3 = 1.975$, $x_4 = 2.675$.

The corresponding dynamic equations (2) are then

$$\frac{dx_i}{dt} = -a_{i\alpha}x_\alpha + c_i, \qquad i = 1, \cdots, 4.$$

The computation is simplified, as in the relaxation method, if one first determines the change b_{ij} in the derivative $\dot{x}_i = dx_i/dt$ corresponding to a unit change in each x_j. Here one has simply $b_{ij} = -a_{ij}$. All changes in the derivatives can thus be read off the matrix of the a_{ij}.

If one starts at $(x_1^0, x_2^0, x_3^0, x_4^0) = (0, 0, 0, 0)$, then the computation is as shown in Table 2. The relaxation procedure can be summarized in a similar fashion, shown in Table 3. In this example the new method is clearly faster than relaxation. However, considerable further experimentation is required to determine what is best in general.

TABLE 1

x	y	F	$\partial F/\partial n$	x	y	F	$\partial F/\partial n$
0	0	0	—	3	3	4	—
1	0	1	0	2	3	2	1
2	0	2	0	1	3	1	1
3	0	3	—	0	3	0	—
3	1	4	1	0	2	1	1
3	2	5	1	0	1	2	1

TABLE 2

x_1	x_2	x_3	x_4	$\dot{x}_1$	$\dot{x}_2$	$\dot{x}_3$	$\dot{x}_4$	e_1	e_2	e_3	e_4	h
0	0	0	0	2	31	15	24	0	1	1	1	1
0	1	1	1	16	15	−1	18	1	1	1	0	1
1	2	1	2	0	9	13	12	0	1	1	0	.1
1	2.1	1.1	2	1.6	6.6	10.6	3.6	0	1	1	0	.1
1	2.2	1.2	2	3.2	4.2	8.2	5.2	1	1	1	1	.1
1.1	2.3	1.3	2.1	2.4	3.4	7.4	4.4	1	1	1	1	.3
1.4	2.6	1.6	2.4	0	1	5	2	0	0	1	0	.1
1.4	2.6	1.7	2.4	.8	.8	2.8	2.8	0	0	1	1	.1
1.4	2.6	1.8	2.5	1.4	1.4	1.4	1.4	1	1	1	1	.2
1.6	2.8	2.0	2.7	−.2	−.2	−.2	−.2					

TABLE 3

x_1	x_2	x_3	x_4	$\dot{x}_1$	$\dot{x}_2$	$\dot{x}_3$	$\dot{x}_4$	e_1	e_2	e_3	e_4	h
0	0	0	0	2	31	15	24	0	1	0	0	1
0	1	0	0	10	9	13	32	0	0	0	1	1
0	1	0	1	8	17	21	10	0	0	1	0	1
0	1	1	1	16	15	−1	18	0	0	0	1	1
0	1	1	2	14	23	7	−4	0	1	0	0	1
0	2	1	2	22	1	5	4	1	0	0	0	1
1	2	1	2	0	9	13	2	0	0	1	0	.6
1	2	1.6	2	4.8	7.8	−.2	0.8	0	1	0	0	.4
1	2.4	1.6	2	8.0	−1.0	−1.0	10.0	0	0	0	1	.5
1	2.4	1.6	2.5	7.0	3.0	3.0	−1.0	1	0	0	0	.3
1.3	2.4	1.6	2.5	.4	5.4	5.4	−1.6	0	1	0	0	.2
1.3	2.6	1.6	2.5	2	1.0	5.0	0	0	0	1	0	.2
1.3	2.6	1.8	2.5	3.6	.6	0.6	1.6	1	0	0	0	.2
1.5	2.6	1.8	2.5	−.8	2.2	2.2	1.2	0	1	0	0	.1
1.5	2.7	1.8	2.5	0	0	2.0	2.0	0	0	0	1	.1
1.5	2.7	1.8	2.6	−.2	.8	2.8	−.2	0	0	1	0	.1
1.5	2.7	1.9	2.6	.6	.6	.6	.6					

UNIVERSITY OF MICHIGAN,
ANN ARBOR, MICH.

LARGE DEFLECTION THEORY FOR RECTANGULAR PLATES

BY

SAMUEL LEVY

1. Introduction. During flight, the metal covering or "skin" of modern aircraft is relied upon to carry a large proportion of the load on the airplane and, at the same time, to be aerodynamically "smooth" in order to prevent excessive drag. An analysis of thin plates by the large deflection theory is essential for a fundamental understanding of the load-carrying capacity and strength after buckling of this metal skin.

The skin is generally divided into rectangular fields by the stiffeners, ribs, bulkheads or other structural members. Each field can be considered as a rectangular elastic plate supported along the edges and subjected to edge loads and normal loads. These loads approximate the loads imposed on the aircraft structure during various critical flight conditions. The analysis of such loaded plates is complicated by the fact that the deflections of the plate may be comparable in magnitude with the plate thickness. In such cases Kirchhoff's linear plate theory may yield results that are considerably in error and a more rigorous theory that takes account of deformations in the middle surface should be applied.

The fundamental non-linear large deflection equations for the more exact theory were derived by von Kármán about 35 years ago. A procedure for solving these equations has been worked out at the National Bureau of Standards for the National Advisory Committee for Aeronautics. This paper presents this procedure and some of the results obtained by using it.

2. Fundamental equations.

2.1. *Symbols.* An initially flat rectangular plate of uniform thickness will be considered. The symbols have the following significance:

- a plate length in x-direction
- b plate length in y-direction
- h plate thickness
- x, y coordinate axes with origin at corner of plate
- w vertical displacement of points of the middle surface
- p lateral pressure
- F the Airy stress function
- E Young's modulus
- μ Poisson's ratio
- $D = Eh^3/12\,(1 - \mu^2)$, flexural rigidity of plate

Subscripts m, n, p, q, r, s represent integers.

2.2. *Plate equations.* Von Kármán's equations for the deflection w and Airy's stress function F for thin flat plates are [9, p. 323]:

$$(1)\qquad \frac{\partial^4 F}{\partial x^4} + 2\frac{\partial^4 F}{\partial x^2\,\partial y^2} + \frac{\partial^4 F}{\partial y^4} = E\left[\left(\frac{\partial^2 w}{\partial x\partial y}\right)^2 - \frac{\partial^2 w}{\partial x^2}\frac{\partial^2 w}{\partial y^2}\right],$$

$$(2)\qquad \frac{\partial^4 w}{\partial x^4} + 2\frac{\partial^4 w}{\partial x^2\,\partial y^2} + \frac{\partial^4 w}{\partial y^4} = \frac{p}{D} + \frac{h}{D}\left(\frac{\partial^2 F}{\partial y^2}\frac{\partial^2 w}{\partial x^2} + \frac{\partial^2 F}{\partial x^2}\frac{\partial^2 w}{\partial y^2} - 2\frac{\partial^2 F}{\partial x\partial y}\frac{\partial^2 w}{\partial x\partial y}\right).$$

2.3. *Boundary conditions.* A solution of von Kármán's equations is required for a variety of boundary conditions at the four edges of the plates. These boundary conditions depend on the degree of fixity at the edges of the plate and on the loads applied to those edges when the plate is considered a part of the structure of the airplane.

There are boundary conditions for both the lateral displacement w and the stress function F.

Simply supported edges. In this case both the deflections and the bending moments along the edges are zero, or

$$(3)\qquad \begin{aligned} (w)_{x=0,a} &= 0, & (w)_{y=0,b} &= 0,\\ \left(\frac{\partial^2 w}{\partial x^2}\right)_{x=0,a} &= 0, & \left(\frac{\partial^2 w}{\partial y^2}\right)_{y=0,b} &= 0. \end{aligned}$$

Built-in edges. In this case both the deflections and slopes along the edges are zero, or

$$(4)\qquad \begin{aligned} (w)_{x=0,a} &= 0, & (w)_{y=0,b} &= 0,\\ \left(\frac{\partial w}{\partial x}\right)_{x=0,a} &= 0, & \left(\frac{\partial w}{\partial y}\right)_{y=0,b} &= 0. \end{aligned}$$

Straight edges. If the elongation of the plate is δ_x in the x-direction and δ_y in the y-direction, and if the edges of the plate are straight, then δ_x and δ_y must be independent of the position along the length or width at which they are measured, or

$$(5)\qquad \begin{aligned} \int_0^a \left[\frac{1}{E}\left(\frac{\partial^2 F}{\partial y^2} - \mu\frac{\partial^2 F}{\partial x^2}\right) - \frac{1}{2}\left(\frac{\partial w}{\partial x}\right)^2\right] dx &= \delta_x,\\ \int_0^b \left[\frac{1}{E}\left(\frac{\partial^2 F}{\partial x^2} - \mu\frac{\partial^2 F}{\partial y^2}\right) - \frac{1}{2}\left(\frac{\partial w}{\partial y}\right)^2\right] dy &= \delta_x. \end{aligned}$$

Loads on the edges. If the load in the plane of the plate is P_x in the x-direction and P_y in the y-direction,

$$(6)\qquad h\left[\left(\frac{\partial F}{\partial y}\right)_{y=b} - \left(\frac{\partial F}{\partial y}\right)_{y=0}\right] = P_x, \qquad h\left[\left(\frac{\partial F}{\partial x}\right)_{x=a} - \left(\frac{\partial F}{\partial x}\right)_{x=0}\right] = P_y.$$

3. **General solution.** A general solution of the plate equations (1) and (2) in terms of trigonometric series was suggested by the experimentally observed deflected shape of plates under load. This solution is developed in [1].

3.1. *Lateral deflection.* The lateral deflection was taken in [1] as

$$w = \sum_{m=1}^{\infty} \sum_{n=1}^{\infty} w_{mn} \sin \frac{m\pi x}{a} \sin \frac{n\pi y}{b}. \tag{7}$$

Equation (7) satisfied the boundary conditions (3) for simply-supported edges automatically. In [3] it is shown that equation (7) will also satisfy the boundary conditions, equations (4), for built-in edges if the coefficients w_{mn} satisfy the equations

$$\begin{aligned} \sum_{m=1}^{\infty} m w_{mn} &= 0, & \sum_{m=1}^{\infty} (-1)^m m w_{mn} &= 0, \\ \sum_{n=1}^{\infty} n w_{mn} &= 0, & \sum_{n=1}^{\infty} (-1)^n n w_{mn} &= 0. \end{aligned} \tag{8}$$

3.2. *Lateral pressure.* The lateral pressure p distributed over the surface of the plate was considered to consist of two parts p' and p''. The part p'' corresponded with lateral pressures near the edges of the plate whose purpose was to apply edge moments in those cases where the edges were built-in. The part p' corresponded with the given pressure distribution on all parts of the plate except the boundary.

The pressure p was expressed as a Fourier series,

$$p = \sum_{m=1}^{\infty} \sum_{n=1}^{\infty} p_{mn} \sin \frac{m\pi x}{a} \sin \frac{n\pi y}{b}, \tag{9}$$

where

$$p_{mn} = q'_{mn} + q''_{mn}. \tag{9a}$$

The coefficients q'_{mn} were determined from the given pressure distribution p' by:

$$q'_{mn} = \frac{4}{ab} \int_0^a \int_0^b p' \sin \frac{m\pi x}{a} \sin \frac{n\pi y}{b} \, dx \, dy. \tag{10}$$

The coefficients q''_{mn} were determined from the pressure distribution p'' near the edges of the plate by a method similar to that in [3]. It was found that:

$$q''_{mn} = n[k'_m - (-1)^n k''_m] + m[t'_n - (-1)^m t''_n] \tag{11}$$

where the edge moments were positive if they deflected the plate positively and were given by:

$$\begin{aligned} &\sum_{m=1}^{\infty} \frac{b^2}{2\pi} k'_m \sin \frac{m\pi x}{a}, && \text{along edge } y = 0, \\ &\sum_{m=1}^{\infty} \frac{b^2}{2\pi} k''_m \sin \frac{m\pi x}{a}, && \text{along edge } y = b, \\ &\sum_{n=1}^{\infty} \frac{a^2}{2\pi} t'_n \sin \frac{n\pi y}{b}, && \text{along edge } x = 0, \\ &\sum_{n=1}^{\infty} \frac{a^2}{2\pi} t''_n \sin \frac{n\pi y}{b}, && \text{along edge } x = a. \end{aligned} \tag{12}$$

3.3. *Airy stress function F.* In order to satisfy equation (1) and boundary condition equation (6), F was taken as:

$$F = \frac{P_x y^2}{2bh} + \frac{P_y x^2}{2ah} + \sum_{m=0}^{\infty} \sum_{n=0}^{\infty} f_{mn} \cos \frac{m\pi x}{a} \cos \frac{n\pi y}{b}. \tag{13}$$

Substituting equations (13) and (7) simultaneously into equation (1), and equating coefficients of like trigonometric terms on the two sides of the equation, results in the relation between the coefficients f_{mn} and w_{mn} given in [1]. This relation can be simplified to:

$$f_{mn} = \frac{E}{4(m^2 b/a + n^2 a/b)^2} \sum b_{rspq} w_{rs} w_{pq} \tag{14}$$

where:

(1) The summation includes all products $w_{rs}\, w_{pq}$ for which m is equal to either the sum or difference of r and p, and for which n is equal to either the sum or difference of s and q.

(2) The coefficient b_{rspq} is given by:

$$b_{rspq} = 2\, rspq \pm (r^2 q^2 + s^2 p^2) \tag{14a}$$

where the sign before the parenthesis is positive if m is the sum of r and p, and n is the difference of s and q; or if m is the difference of r and p, and n is the sum of s and q. It is negative otherwise.

An example of the use of equation (14) for the case of a square plate ($a = b$) is

$$f_{2,4} = \frac{E}{1600} (-4w_{1,1} w_{1,3} + 36 w_{1,1} w_{3,3} + 36 w_{1,1} w_{1,5} + 64 w_{1,2} w_{1,6} \cdots). \tag{14b}$$

3.4. *Equilibrium equation.* It remains to show the necessary relation between the coefficients w_{mn}, equation (7), p_{mn}, equation (9), and f_{mn}, equation (14). Substituting equations (7), (9), and (13) simultaneously into the lateral equilibrium equation, equation (2), and equating coefficients of like trigonometric terms on the two sides of the equation results in the relation given in equation (9) of [1]. This relation can be simplified to:

$$\begin{aligned} p_{mn} &= D w_{mn} \pi^4 \left(\frac{m^2}{a^2} + \frac{n^2}{b^2}\right)^2 + P_x w_{mn} \frac{m^2 \pi^2}{a^2 b} \\ &\quad + P_y w_{mn} \frac{n^2 \pi^2}{ab^2} + \frac{h\pi^4}{4a^2 b^2} \sum d_{rspq} f_{rs} w_{pq} \end{aligned} \tag{15}$$

where:

(1) The summation includes all products $f_{rs} w_{pq}$ for which m is equal to either the sum or difference of r and p, and for which n is equal to either the sum or difference of s and q.

(2) The coefficients d_{rspq} are given by:

$$d_{rspq} = \pm (rq \pm sp)^2 \text{ if } r \neq 0,\ s \neq 0 \tag{15a}$$

and are twice this value if either r or s is zero. The signs in equation (15a) are determined by the following rules: The first sign is positive if either $r - p = m$ or if $s - q = n$ but not if both conditions are true. It is negative in all other cases. The second sign is positive if m is the sum of r and p, and n is the difference of s and q; or if m is the difference of r and p, and n is the sum of s and q. It is negative otherwise.

An example of the use of equation (15) for the case $m = 1, n = 3$ is:

$$p_{1,3} = Dw_{1,3}\pi^4\left(\frac{1}{a^2} + \frac{9}{b^2}\right)^2 + P_x w_{1,3}\frac{\pi^2}{a^2 b} + P_y w_{1,3}\frac{9\pi^2}{ab^2} \tag{15b}$$
$$+ \frac{h\pi^4}{4a^2b^2}(-8f_{0,2}w_{1,1} - 8f_{0,2}w_{1,5} + 100f_{2,4}w_{3,1} - 64f_{2,2}w_{3,1} + \cdots).$$

3.5. *Straight edges.* The boundary condition for straight edges, equation (5), can be shown to be satisfied by the chosen expressions for F and w by substituting equations (7) and (13) into equations (5), and performing the indicated differentiation and integration. After considerable simplification, requiring the use of relation (14), it is found that the edges $x = 0$ and $x = a$ separate by an amount

$$\delta_x = \frac{P_x a}{bhE} - \mu\frac{P_y}{hE} - \frac{\pi^2}{8a}\sum_{m=1}^{\infty}\sum_{n=1}^{\infty} m^2 w_{mn}^2, \tag{16a}$$

while the edges $y = 0$ and $y = b$ separate by an amount

$$\delta_y = -\mu\frac{P_x}{hE} + \frac{P_y b}{ahE} - \frac{\pi^2}{8b}\sum_{m=1}^{\infty}\sum_{n=1}^{\infty} n^2 w_{mn}^2. \tag{16b}$$

Since the right-hand sides of equations (16a) and (16b) are independent of x and y, the boundary conditions for straight edges are satisfied.

3.6. *Finite set of equations.* Equations (8), (14), and (15) each represent infinite sets of equations. An exact solution of a given problem should satisfy all of these equations. A good approximate solution can, however, be obtained by assuming that only a limited number of the w_{mn} coefficients differ from zero and that, of these, only a few will have large enough values so that they cannot be neglected where cubic products occur.

The accuracy of any given solution can be judged by observing the change in the answer as the number of w_{mn} coefficients is gradually increased. In one case [1] a single term was found to give answers accurate to three significant figures. To obtain answers accurate to three significant figures in another case [3], it was necessary to consider 36 w_{mn} coefficients different from zero of which 6 were large enough so that their cubic products could not be neglected.

3.7. *Combination of equations.* In any given problem the unknown coefficients w_{mn}, f_{mn} and, in the case of clamped edges, k'_m, k''_m, t'_n, t''_n, must be determined from the given loading and boundary conditions. A sufficient number of simultaneous equations (equations (15), (14), and (8) respectively) are available for such a solution. The work required to obtain a solution, however, can be considerably reduced if the equations are first combined.

One such combination is to substitute the value of f_{mn} obtained from equation (14) into the equations determined from equation (15) as is done in [1]. When this is done, there results a set of simultaneous equations involving cubic products of the w_{mn} coefficients. In the case of simply-supported edges where k'_m, k''_m, t'_n, t''_n are zero, a solution is obtained by solving these equations as described in the next section.

For the case of plates with clamped edges where k'_m, k''_m, t'_n, and t''_n are not zero, an additional useful combination of equations can be made (see [3]). First rearrange the cubic equations of the preceding paragraph so that the linear w_{mn} term is on the left side and all the remaining terms are on the right. Then substitute those equations for the w_{mn} terms in equations (8). The result will be a set of equations involving the k'_m, k''_m, t'_n, and t''_n coefficients linearly and the w_{mn} coefficients as cubic products. In combining these equations, it should be noted that, although it is usually sufficient to include the cubic terms involving only the more important deflection coefficients w_{mn}, it is necessary to include the linear w_{mn} terms from a large number of equations in order to get good accuracy.

3.8. *Solving simultaneous cubic equations.* The combinations of equations just described result in a set of simultaneous cubic equations which must be solved to obtain the stress distribution and deflection of the plate. Various methods have been suggested in the literature from time to time for solving such equations. The one found adequate for the problems investigated by the author is essentially Newton's method and was carried out as follows:

The cubic equations were expanded in a Taylor series in the neighborhood of an estimated solution omitting terms in the Taylor series of higher order than the first. The resulting linear equations were solved for the first order correction to the estimated solution, using Crout's method, and the process was repeated if necessary. One or two repetitions were usually sufficient.

3.9. *Convergence.* The convergence of the solution is assured by the rapid increase in the coefficients d_{rspq} (15a) and b_{rspq} (14a) as r, s, p, and q increase. Some question can be raised in the case of plates with clamped edges since q''_{mn}, equation (11), also increases with m and n. The results obtained indicate that convergence is obtained in spite of this, probably due to the fact that the coefficients q''_{mn} increase so much more slowly than the coefficients d_{rspq} and b_{rspq}.

3.10. *Multiple solutions—stability.* The non-linear character of plate problems makes possible, under certain loading conditions, more than one stable solution. The plate can, in such cases, "oil-can" from one stable configuration to another. In the case of plates subjected to both lateral pressure and edge compression, as the load is gradually increased, a load may be reached at which a given configuration of the plate will cease to be stable and the plate will "snap" to a different configuration. This stability problem, although it is not clearly understood, depends upon whether the edge compression or the over-all shortening is taken as the independent variable of the problem. The physical equivalent of the mathematical use of the edge compression as independent variable is loading in a dead-weight testing machine, while the physical equivalent of the mathematical use of over-all shortening as independent variable is approached in conventional hydraulic and screw-type testing machines. A thorough discussion of this is given by Tsien in [10].

4. **Results.** The method presented in the previous section for solving von Kármán's equations for the behavior of plates under normal pressure and load in the plane of the plate has been applied to the solution of a variety of prob-

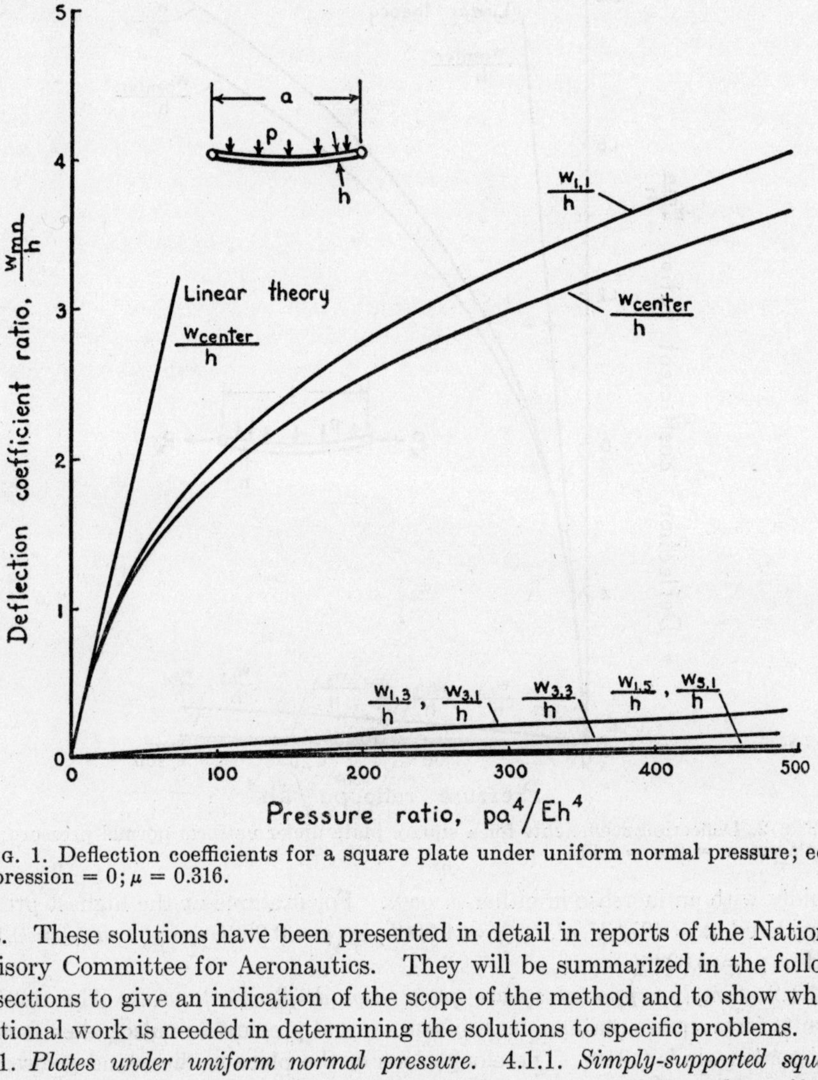

FIG. 1. Deflection coefficients for a square plate under uniform normal pressure; edge compression = 0; $\mu = 0.316$.

lems. These solutions have been presented in detail in reports of the National Advisory Committee for Aeronautics. They will be summarized in the following sections to give an indication of the scope of the method and to show where additional work is needed in determining the solutions to specific problems.

4.1. *Plates under uniform normal pressure.* 4.1.1. *Simply-supported square plate, edge tension zero.* The following results for a square plate under uniform normal pressure are taken from [1]. Poisson's ratio is taken as 0.316. The edge tensions P_x in the x-direction and P_y in the y-direction are zero. The uniform normal pressure is p. The length of the plate edge is a, and the thickness of the plate is h. The computed values of the deflection coefficients, w_{mn}, and of the center deflection, w_{center}, are given in dimensionless form in Fig. 1 which corresponds with Fig. 7 of [1]. It is seen from Fig. 1 that for center deflections greater than about one-half plate thickness the relation between center deflection and pressure becomes non-linear. It is also seen that the w_{mn} coefficients decrease

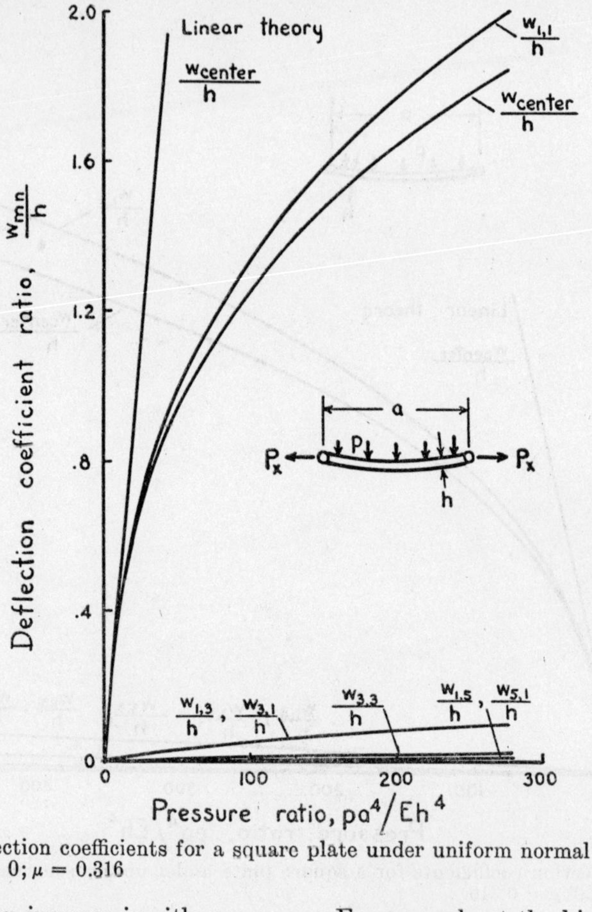

FIG. 2. Deflection coefficients for a square plate under uniform normal pressure; edge displacement = 0; $\mu = 0.316$

rapidly with an increase in either m or n. For example at the highest pressure considered, $p = 497Eh^4/a^4$, $w_{1,1} = 4.000h$, $w_{1,3} = 0.282h = w_{3,1}$, $w_{3,3} = 0.116h$, and $w_{1,5} = 0.0236h = w_{5,1}$.

4.1.2. *Simply-supported square plate, edge displacment zero.* The following results for a square plate under uniform normal pressure are also taken from [1]. Poisson's ratio is 0.316. The elongations of the plate in the x and y directions given by equation (5) are taken as zero. The computed values of the deflection coefficients, w_{mn}, and of the center deflection, w_{center}, are given in dimensionless form in Fig. 2 which corresponds with Fig. 11 of [1]. In this case it is seen that the center deflection-pressure relation becomes non-linear even earlier than was found to be the case in the previous example where the edge compression was taken as zero. At $w_{\text{center}} = 0.3h$ non-linearity is already present.

Convergence for this case is seen to be rapid, just as for the preceding case. At the highest pressure considered, $p = 278.5Eh^4/a^4$, $w_{1,1} = 2.000h$, $w_{1,3} = 0.0978h = w_{3,1}$, $w_{3,3} = 0.0193h$, and $w_{1,5} = 0.0109h = w_{5,1}$.

Comparison of Figs. 1 and 2 shows that preventing the edges from pulling in causes a substantial increase in the stiffness of the plate in the large deflection

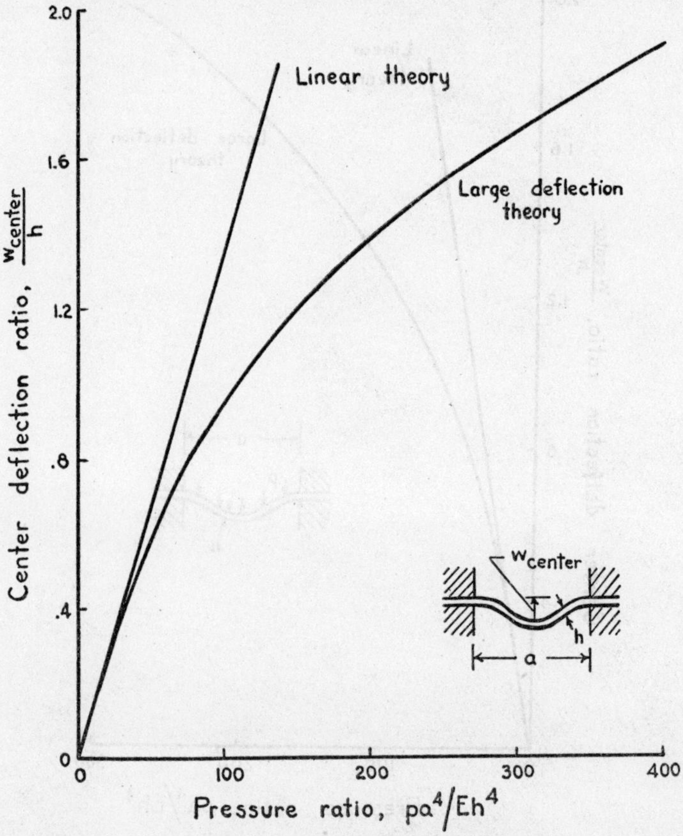

FIG. 3. Center deflection of square plate with clamped edges under uniform norma pressure; $\mu = 0.316$

range. For example, in Fig. 1 to obtain a center deflection of $1.5h$ requires only a pressure $p = 65Eh^4/a^4$, while in Fig. 2 to obtain a center deflection of $1.5h$ requires a pressure $p = 157Eh^4/a^4$, or about 2.4 times as much pressure.

4.1.3. *Clamped square plate.* The following results for a clamped square plate under uniform normal pressure are taken from [3]. Poisson's ratio is 0.316. The computed value of the center deflection is given in dimensionless form in Fig. 3 as a function of the pressure. Non-linearity in this case begins at a center deflection of about $0.3h$. Some idea of the rate of convergence can be obtained by considering the values of the deflection coefficients at the highest pressure considered, $p = 402Eh^4/a^4$. At this pressure, $w_{1,1} = 1.800h$, $w_{1,3} = -0.0854h = w_{3,1}$, $w_{3,3} = 0.0139h$, $w_{1,5} = -0.0681h = w_{5,1}$, and so on. Comparison with the previous cases shows that the convergence is almost as rapid at high pressure as for the plates with simply-supported edges. Solutions of clamped edge plate problems might, therefore, be expected to be as easy as those for simply-supported edges if it were not for the additional simultaneous cubic

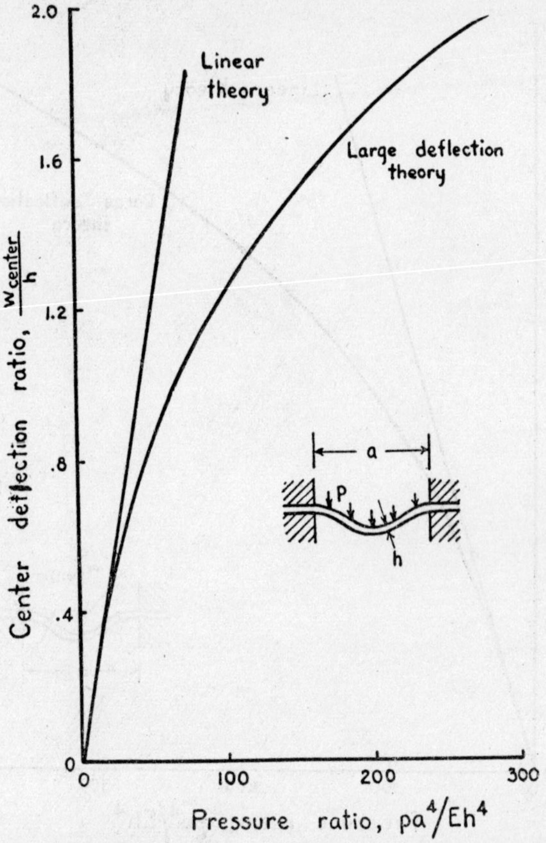

FIG. 4. Center deflection of clamped rectangular plate with length-width ratio of 1.5 under uniform normal pressure; $\mu = 0.316$, width of plate $= a$

equations which must be solved to determine the unknown edge moment coefficients.

It is also of interest to observe the rate of convergence of the edge moment coefficients given in equation (12). Presenting the results of [3] in the notation of the present paper, we find, at the highest pressure considered, $p = 402Eh^4/a^4$, $k_1' = k_1'' = t_1' = t_1'' = -70.0Eh^4/a^4$, $k_3' = k_3'' = t_3' = t_3'' = 0.65Eh^4/a^4$, and so on. It is apparent that the coefficients decrease fairly rapidly as the order of the term increases.

4.1.4 *Clamped rectangular plate.* The following results for a clamped rectangular plate, length-width ratio 1.5 to 1, are taken from [6]. The plate is loaded by a uniform normal pressure p. Poisson's ratio is taken as 0.316. The computed center deflection is given in dimensionless form in Fig. 4 as a function of pressure. Non-linearity again begins at a center deflection of about $0.3h$. In this case, the convergence of the deflection coefficients is about as fast as that for the square plate discussed in the previous section. At a pressure $p = 282Eh^4/a^4$, $w_{1,1} = 2.000h$, $w_{1,3} = 0.075h$, $w_{3,1} = -0.149h$, $w_{5,1} = -0.073h$, $w_{1,5} = -0.064h$, $w_{3,3} = 0.003h$, and so on. The convergence of the edge mo-

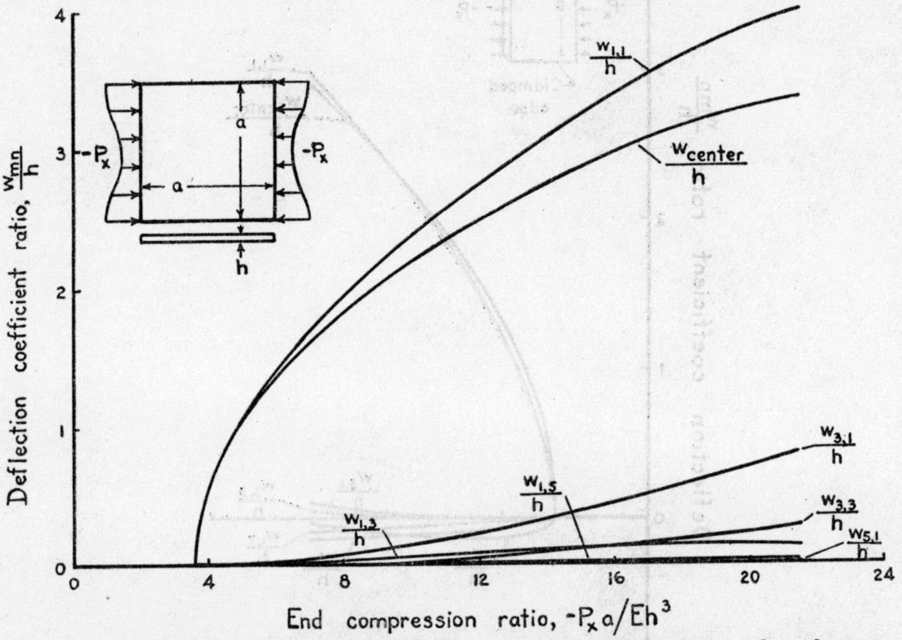

FIG. 5. Deflection coefficients for a simply-supported square plate under edge compression; $\mu = 0.316$

ment coefficients of equation (12) is also fairly rapid. At $p = 282Eh^4/a^4$, $t_1' = t_1'' = -69.0Eh^4/a^4$, $k_1' = k_1'' = -24.0Eh^4/a^4$, $t_3' = t_3'' = -4.7Eh^4/a^4$, $k_3' = k_3'' = 0.6Eh^4/a^4$, and so on.

4.2. *Plates under edge compression in one direction.* 4.2.1. *Simply-supported square plate.* The following results taken from [1] apply to simply-supported square plates loaded by edge compression in one direction only. Poisson's ratio is taken as 0.316. The normal pressure p and the edge compression $-P_y$ in the y-direction are zero. The edge compression in the x-direction is $-P_x$. The length of the plate edge is a, and the thickness of the plate is h. The computed values of the deflection coefficients, w_{mn}, and of the center deflection, w_{center}, are given in dimensionless form in Fig. 5 which corresponds with Fig. 2 of [1]. It is seen from Fig. 5 that the center deflection is a non-linear function of the edge compression from the start of buckling. It is also seen that the w_{mn} coefficients decrease rapidly with an increase in either m or n.

4.2.2. *Rectanglar plate clamped along two edges.* The following results, taken from [7], apply to rectangular plates having a length ratio for the two edges of 1 to 1.5. The length of the longer edges is b. The shorter edges are considered clamped and the longer edges simply-supported. The shorter edges are taken

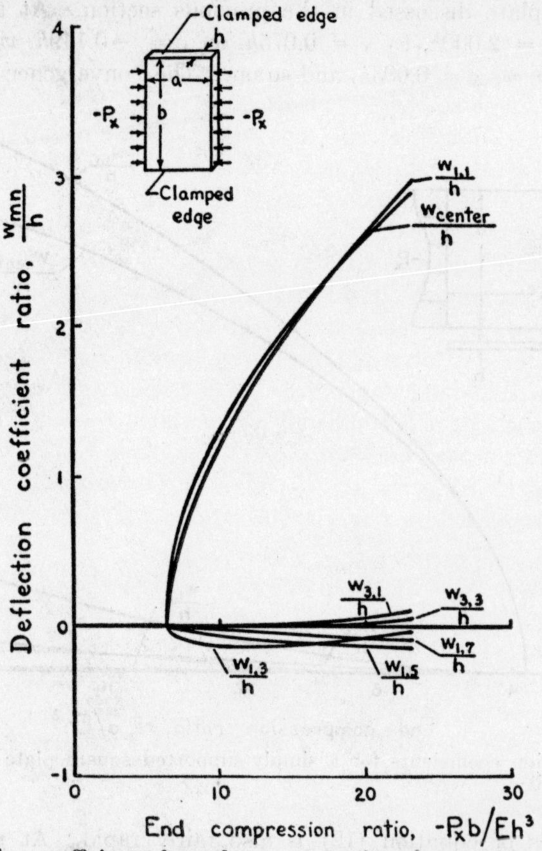

FIG. 6. Deflection coefficients for a long rectangular plate under edge compression. Loaded edges simply-supported, other edges clamped; $\mu = 0.316$

in the x-direction. The normal pressure p and the edge tension in the y-direction, P_y, are taken as zero. The edge compression in the x-direction is $-P_x$. The computed values of the deflection coefficients, w_{mn}, and of the center deflection, w_{center}, are given in dimensionless form in Fig. 6 which corresponds with Fig. 3 of [7]. Again, the center deflection is a non-linear function of the edge compression from the start of buckling and the w_{mn} coefficients decrease rapidly with an increase in either m or n.

4.3. *Plates under combined load, plates under shear, and plates having initial curvature.* Additional solutions of plate problems by the methods presented in this paper or by minor modifications of these methods are presented in [1] and in [5, 11, 4, 8, 2].

The case of combined uniform normal pressure and edge compression in one direction is treated in [1] for a square simply-supported plate and for a rectangular (3 to 1) simply-supported plate. The same combined loading is considered in more detail for a rectangular (4 to 1) plate with simply-supported edges [5] and with clamped edges [11]. The results obtained showed that the non-linearities which occurred were much more pronounced for combined loading than they were for a single loading.

The cases of simply-supported plates under shear are treated in [4] and [8]. [4] presents results for a shear web divided into square plates by the reinforcing struts; while, [8] presents similar results for a shear web divided into rectangular (2.5 to 1) plates by the reinforcing struts.

An extension of the methods of the present paper to plates having a moderate amount of initial curvature to form a cylindrical surface is presented in [2]. Numerical results are presented for plates whose initial shape is that of a circular cylinder. The loading is taken as an edge compression in the direction of the generators. In these solutions, just as for the solutions for plates under combined normal pressure and edge compression, the non-linearities were extreme.

5. Recommendations for additional investigation. Proof of the validity of some of the mathematical methods used in this paper has not been given. Although results obtained by the method are in agreement with past experience, in order to put the method on a sound footing, it would be desirable to clarify the following:

(1) The pressure distribution in equations (9) and (11), which replaces the edge moments in the case of clamped edge plates, has coefficients q''_{mn} which increase indefinitely as m or n increases. This nonconvergent pressure distribution is used as though it were convergent and the resulting values of w_{mn} appear to converge. It would be desirable to prove that the values of w_{mn} must converge.

(2) Because of the non-linearity of the equations, more than one solution can sometimes be found. A criterion by which one could determine whether a given solution corresponds to stable or unstable equilibrium would be of value in buckling problems.

Additional investigation to extend the numerical results already available would be useful. It would be particularly desirable to extend the results for plates under normal pressure to include a few additional length to width ratios for simply-supported plates and to consider simply-supported plates under combined axial stress in both the x and y directions.

Bibliography

1. Samuel Levy, *Bending of rectangular plates with large deflections*, NACA Report No. 737 and NACA T.N. No. 846, 1942.

2. ———, *Large deflection theory of curved sheet*, NACA T.N. No. 895, 1943.

3. ———, *Square plate with clamped edges under normal pressure producing large deflections*, NACA Report No. 740 and NACA T.N. No. 847, 1942.

4. Samuel Levy, Kenneth L. Fienup, and Ruth M. Woolley, *Analysis of square shear web above buckling load*, NACA T.N. No. 962, 1945.

5. Samuel Levy, Daniel Goldenberg, and George Zibritosky, *Simply-supported long rectangular plate under combined axial load and normal pressure*, NACA T.N. No. 949, 1944.

6. Samuel Levy, and Samuel Greenman, *Bending with large deflection of a clamped rectangular plate with length-width ratio of* 1.5 *under normal pressure*, NACA T.N. No. 853, 1942.

7. Samuel Levy and Philip Krupen, *Large deflection theory for end compression of long rectangular plates rigidly clamped along two edges*, NACA T.N. No. 884, 1943.

8. Samuel Levy, Ruth M. Woolley and Josephine N. Corrick, *Analysis of deep rectangular shear web above buckling load*, NACA T.N. No. 1009, 1946.

9. S. Timoshenko, *Theory of elastic stability*, New York, McGraw-Hill, 1936.

10. Hsue-Shen Tsien, *A theory for the buckling of thin shells*, Journal of Aeronautical Sciences vol. 9 (1942) pp. 373–384.

11. Ruth M. Woolley, Josephine N. Corrick, and Samuel Levy, *Clamped long rectangular plate under combined axial load and normal pressure*, NACA T.N. No. 1047, 1946.

NATIONAL BUREAU OF STANDARDS,
WASHINGTON, D.C.

DISCONTINUOUS SOLUTIONS IN THE THEORY OF PLASTICITY[1]

BY

WILLIAM PRAGER

According to Saint Venant's theory of plasticity the stress distribution becomes statically determinate when the plastic medium is in a state of plane flow under the action of given boundary stresses. These statically determinate problems of plasticity have formed the subject of many investigations.

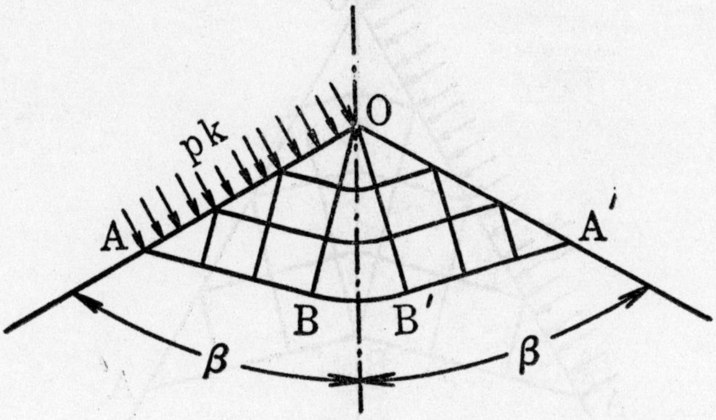

FIG. 1

The partial differential equations which govern these statically determinate stress distributions allow discontinuities of the first derivatives of the stress components across the lines of maximum shearing stress (characteristics). Since these discontinuities obviously correspond to the "sonic discontinuities" in the dynamics of compressible fluids, it seems natural to look for the analogue, in plasticity, of the stronger discontinuities (shock fronts, contact surfaces) encountered in the dynamics of compressible fluids.

That such stronger discontinuities are indeed possible in plasticity can be shown as follows. Let N and T denote the normal and tangential stress transmitted across a cylindrical discontinuity surface S at a generic point P of this surface. Equilibrium at P requires continuity of N and T across the surface of discontinuity. It can be shown that the necessary continuity of the velocity components does not furnish any additional continuity condition for the stress components. The normal stress N' transmitted across a surface element which is parallel to the generators of S, but normal to S, may therefore be discontinuous. Since the stress components must satisfy the yield condition

[1] The full paper appeared in the R. Courant Anniversary Volume, Interscience Publishers, 1948. The results presented in this paper were obtained in the course of research conducted under Contract N7onr-358 sponsored jointly by the Office of Naval Research and the Bureau of Ships.

$$(N - N')^2 + 4T^2 = 4k^2$$

(k = yield stress in pure shear) on both sides of S, we have

$$N' = N \pm 2(k^2 - T^2)^{1/2}.$$

The stress component N' can therefore jump by $\pm 4(k^2 - T^2)^{1/2}$ as one passes from one side of S to the other.

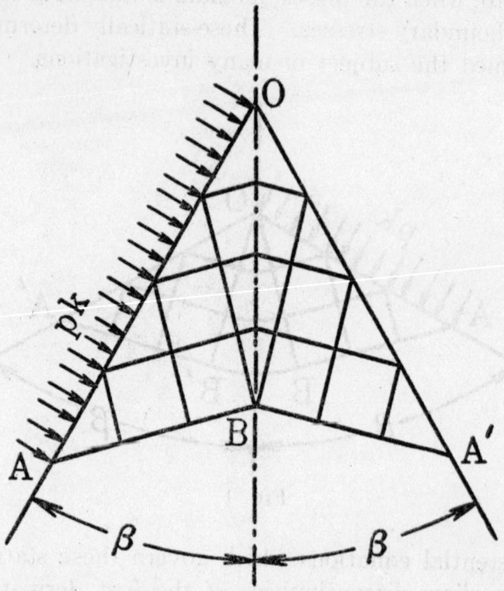

FIG. 2

The following example indicates that discontinuities of this kind are bound to arise in certain two-dimensional problems of plasticity. Fig. 1 shows the well known pattern of shear lines for a wedge AOA' which is subject to uniformly distributed normal pressure pk along the portion AO of one of its sides. The pressure necessary to produce the plastic zone $OABB'A'$ corresponds to

$$p = 2 + 4\beta - \pi,$$

where 2β is the wedge angle. The isosceles right triangles AOB and $A'OB'$ are regions of constant state. This continuous pattern of shear lines occurs only if $\beta > \pi/4$. For $\beta < \pi/4$ the isosceles right triangles AOB and $A'OB'$ would overlap, and this is obviously not admissible. For $\beta < \pi/4$ the regions of constant state adjacent to AO and $A'O$, respectively, are therefore separated by a discontinuity surface. It is found that this discontinuity surface bisects the wedge angle (Fig. 2).

BROWN UNIVERSITY,
PROVIDENCE, R. I.

ON FINITE DEFLECTIONS OF CIRCULAR PLATES

BY

ERIC REISSNER

1. Introduction. In this note we consider the problem of rotationally symmetric deformations of thin, circular, elastic plates. We assume small strains but make no restrictions with regard to the magnitude of displacements and slopes. Our problem is thus more general than would follow from the known theory of plates with small finite deflections where it is assumed that the squares of the slope of the deflected middle surface are of the same order of magnitude as the elastic strains.[1]

Due to the deformation the circular plate becomes a shell of revolution. Consequently our developments are similar in many respects to the method which H. Reissner [6] has given for the analysis of spherical shells and which E. Meissner [5] has shown to be applicable to general shells of revolution.

In what follows we obtain a system of two simultaneous differential equations for the slope ϕ of the deflected middle surface and for the horizontal stress resultant H of the deflected plate. If triple and higher order products of ϕ and H are omitted and simple products are retained our equations reduce to those which have formed the basis of the earlier work on small finite deflections of circular plates by Hencky, Federhofer, Nadai, Prescott, Way, and Friedrichs and Stoker [2].

In analogy to the situation in the small deflections theory of shells of revolution we have for the circular plate subjected to edge loads and edge moments the possibility of boundary layer effects [2]. By boundary layer effect (or edge effect) is meant the fact that a significant redistribution of the edge stresses occurs in a zone adjacent to the boundary, the width of this zone being small compared with the radius of the plate. In the present paper we obtain a system of equations for this effect, which generalizes the result of Friedrichs and Stoker [2] for the case of small finite deformations to the case of arbitrarily large deformations (which are compatible with small strains if the plate is sufficiently thin).

The boundary layer equations of this paper may be solved by the same series method which was used by Friedrichs and Stoker in their work [2]. Solutions of specific problems on the basis of the present result will be given in a subsequent paper.

2. Geometry and strains of the deflected plate. We choose as coordinates on the deflected middle surface of the plate the quantities r and θ where r is the

---

[1] It appears that this theory has first been formulated by G. Kirchhoff [4], essentially in the form of three simultaneous differential equations for the components u, v, w of the displacements of the middle surface of the plate. Instead of three equations for u, v, and w it is possible (in problems in which $\partial/\partial t = 0$) to obtain two simultaneous equations for w and Airy's stress function F. This formulation of the theory has been given by Th. von Kármán [3].

distance from the axis of revolution of a point on the deflected middle surface before deformation and where θ is the circumferential angle. A point which on the undeflected middle surface had the coordinates r, θ and $z = 0$ has on the deflected middle surface the coordinates $r + u(r)$, θ and $z = w(r)$. The quantities u and w are components of radial and axial displacement, respectively.

The deflected middle surface has the linear element $ds^2 = \alpha_r^2 dr^2 + \alpha_\theta^2 d\theta^2$, where the coefficients of the linear element are given by the formulas

$$\text{(1a)} \qquad \alpha_r = [(1 + u')^2 + (w')^2]^{1/2} = 1 + \epsilon_r,$$

$$\text{(1b)} \qquad \alpha_\theta = r + u = r(1 + \epsilon_\theta).$$

In equations (1) and in what follows primes denote differentiation with respect to r and ϵ_r and ϵ_θ are the components of strain of the middle surface. We have then, for the first of the two following equations, in view of the assumption of small strain,

$$\text{(2a)} \qquad \epsilon_r = u' + [(u')^2 + (w')^2]/2,$$

$$\text{(2b)} \qquad \epsilon_\theta = u/r.$$

Note that as we wish to admit the possibility of large displacements we do not neglect the term $(u')^2$ compared with $(w')^2$.

We further need the principal radii of the curvature of the deformed middle surface. For these we have the formulas

$$\text{(3a)} \qquad \frac{1}{R_\theta} = \frac{z'}{(r + u)\alpha_r} = \frac{w'}{r}\,\frac{1}{(1 + \epsilon_r)(1 + \epsilon_\theta)},$$

$$\text{(3b)} \qquad \frac{1}{R_r} = \frac{(r' + u')z'' - (r'' + u'')z'}{\alpha_r^3} = \frac{w'' + u'w'' - u''w'}{(1 + \epsilon_r)^3}.$$

Equations (2) and (3) assume a form which is more convenient for what follows if we introduce as one dependent variable the angle ϕ between the r-direction and the radial tangent to the deformed middle surface. We then have instead of (2a)

$$\text{(4a)} \qquad \epsilon_r = \frac{1 + u'}{\cos \phi} - 1$$

with ϵ_θ as before given by (2b).

Equations (3) become

$$\text{(5a)} \qquad 1/R_\theta = \sin \phi / r,$$

$$\text{(5b)} \qquad 1/R_r = d\phi/ds = \phi'/(1 + \epsilon_r).$$

Finally, equations (4a) and (2b) imply as compatibility equation

$$\text{(6)} \qquad (r\epsilon_\theta)' - \cos \phi \, \epsilon_r = \cos \phi - 1.$$

We now proceed to the calculation of the components of strain at points away from the middle surface of the deformed plate. We neglect the effect of transverse shear and transverse normal stress deformation and designate the components of strain at a point with distance ζ from the middle surface by $\epsilon_r^{(\zeta)}$ and $\epsilon_\theta^{(\zeta)}$. It may be seen by means of a simple diagram that the following relations hold

(7a, b) $$\epsilon_\theta^{(\zeta)} = \frac{2\pi(r + u - \zeta \sin \phi) - 2\pi r}{2\pi r}, \qquad \frac{1 + \epsilon_r^{(\zeta)}}{1 + \epsilon_r} = \frac{R_r - \zeta}{R_r}.$$

Equations (7) may be written in the form

(8a, b) $$\epsilon_\theta^{(\zeta)} = \epsilon_\theta + \zeta\kappa_\theta, \qquad \epsilon_r^{(\zeta)} = \epsilon_r + \zeta\kappa_r$$

where ϵ_θ and ϵ_r are given by (2b) and (4a) and where the terms κ_θ and κ_r are of the form

(9a) $$\kappa_\theta = -\sin \phi/r = -1/R_\theta,$$

(9b) $$\kappa_r = -d\phi/dr = -(1 + \epsilon_r)/R_r.$$

3. Equilibrium equations and stress strain relations for the deflected plate. We introduce stress resultants N_r, N_θ and Q, and stress couples M_r and M_θ. The quantity Q is directed normal to the deformed middle surface while N_r and N_θ are tangential to the deformed middle surface. The differential equations of equilibrium are those known from the small-deflection theory of shells of revolution, with the difference that now the factors pertaining to the geometry of the shell depend on the deformation variables u and ϕ. We have

(10a) $$(\alpha_\theta N_r)' - \alpha_\theta' N_\theta - \alpha_r\alpha_\theta Q/R_r + \alpha_r\alpha_\theta p = 0,$$

(10b) $$(\alpha_\theta Q)' + \alpha_r\alpha_\theta[(N_r/R_r) + (N_\theta/R_\theta)] + \alpha_r\alpha_\theta q = 0,$$

(10c) $$(\alpha_\theta M_r)' - \alpha_\theta' M_\theta - \alpha_r\alpha_\theta Q = 0.$$

Equations (10) are simplified by the assumption of small strain which means that according to equation (1) and (5) we may put in (10)

$$\alpha_\theta = r, \qquad \alpha_r = 1, \qquad 1/R_r = \phi', \qquad 1/R_\theta = \sin \phi/r,$$
$$\alpha_\theta' = 1 + u' = (1 + \epsilon_r) \cos \phi = \cos \phi.$$

We have then instead of (10a) to (10c),

(11a) $$(rN_r)' - \cos \phi\, N_\theta - r\phi' Q + rp = 0,$$

(11b) $$(rQ)' + r[\phi' N_r + \sin \phi\, N_\theta/r] + rq = 0,$$

(11c) $$(rM_r)' - \cos \phi\, M_\theta - rQ = 0.$$

The system of equations (11) is to be completed by the relations between stresses and strains. These are, as in the small-deflection theory of shells,

(12a, b) $$Eh\epsilon_r = N_r - \nu N_\theta, \qquad Eh\epsilon_\theta = N_\theta - \nu N_r,$$

(13a, b) $$M_r = D(\kappa_r + \nu\kappa_\theta), \qquad M_\theta = D(\kappa_\theta + \nu\kappa_r).$$

The quantity D is the bending stiffness factor $Eh^3/12(1 - \nu^2)$.

In view of the relatively simple direct derivations we have not attempted to derive the foregoing results from the general work of Synge and Chien [1].

4. **Reduction to two simultaneous equations for ϕ and H.** We proceed as follows. First evaluate the force equilibrium equations (11a) and (11b). Let H and V stand for horizontal and vertical stress resultant, respectively, so that

$$H = N_r \cos\phi - Q \sin\phi, \tag{14a}$$

$$V = N_r \sin\phi + Q \cos\phi. \tag{14b}$$

Evidently, equilibrium in vertical direction requires that

$$rV = -\int_0^r (p \sin\phi + q \cos\phi) r\, dr. \tag{15}$$

From (14) follows

$$N_r = H \cos\phi + V \sin\phi, \tag{16a}$$

$$Q = -H \sin\phi + V \cos\phi. \tag{16b}$$

Having N_r and Q in terms of H and V it remains to obtain a corresponding formula for N_θ. We put (16) into (11a) to get

$$\cos\phi\, N_\theta = (rH \cos\phi + rV \sin\phi)' + (rH \sin\phi - rV \cos\phi)\phi' + rp. \tag{17}$$

This is simplified by setting

$$p_r = p \cos\phi - q \sin\phi \tag{16c}$$

and by taking V from (15), to read

$$N_\theta = (rH)' + rp_r. \tag{16d}$$

We now introduce N_r and N_θ from (16) into the stress strain relations (12) and the result into the compatibility equation (6). This gives as first of the two simultaneous equations for H and ϕ,

$$\begin{aligned} &\left\{\frac{r}{Eh}\left[(rH)' + rp_r - \nu H \cos\phi - \nu V \sin\phi\right]\right\}' \\ &\quad - \frac{\cos\phi}{Eh}\{H \cos\phi + V \sin\phi - \nu(rH)' - \nu r p_r\} + 1 - \cos\phi = 0. \end{aligned} \tag{18}$$

For plates for which Eh is independent of r, equation (18) reduces to

$$\begin{aligned} &(rH)'' + \frac{1}{r}(rH)' - \left[\frac{\cos^2\phi}{r^2} - \nu\frac{\phi' \sin\phi}{r}\right](rH) + \frac{Eh}{r}(1 - \cos\phi) \\ &\quad = \frac{1}{r}[\cos\phi \sin\phi V + \nu(rV \sin\phi)'] - (1 + \nu \cos\phi)\, p_r - \frac{1}{r}(rp_r)'. \end{aligned} \tag{19}$$

A second equation for H and ϕ is obtained by introducing M_r and M_θ from

(13) and Q from (16) into the moment equilibrium equation (11c) as follows

$$(20)\quad \begin{aligned} -[rD(\phi' + \nu \sin \phi/r)]' + \cos \phi\, D[(\sin \phi/r) + \nu\phi'] \\ - r(-H \sin \phi + V \cos \phi) = 0. \end{aligned}$$

For a plate of constant $D = Eh^3/12(1 - \nu^2)$ equation (20) may be written in the following form

$$(21)\quad \phi'' + \frac{1}{r}\phi' - \frac{1}{r^2}\cos \phi \sin \phi = \frac{1}{D}(H \sin \phi - V \cos \phi).$$

5. **The equations for small finite deflections.** We re-derive this known system of equations by introducing into equations (15) to (21) for the trigonometric functions the first terms of their power series and by neglecting all triple and higher products of the dependent variables and their derivatives.

Restricting attention to the case of a concentrated load at the center plus a uniformly distributed load $q = q_0$ we find from (15)

$$(22)\quad V = -P/2\pi r - q_0 r/2.$$

Equations (16) become

$$(23\text{a, b})\quad N_r = H + \phi V, \qquad Q = -\phi H + V,$$

$$(23\text{c})\quad N_\theta = (rH)' - \phi q_0 r.$$

Equations (19) and (21) become

$$(24\text{a})\quad \begin{aligned} (rH)'' + \frac{1}{r}(rH)' - \frac{1}{r^2}(rH) + \frac{Eh}{2r}\phi^2 \\ = \frac{1+\nu}{r}\phi V + \nu(\phi V)' - 2\phi q_0 - r\phi' q_0, \end{aligned}$$

$$(24\text{b})\quad \phi'' + \frac{1}{r}\phi' - \frac{1}{r^2}\phi = \frac{H\phi - V}{D}.$$

6. **Asymptototic solution.** We now restrict attention to plates which are acted upon by edge forces and moments at a boundary $r = a$ which may be an inner boundary or an outer boundary. To fix the ideas we take here a complete plate of radius a so that we have the case of an outer boundary only. We prescribe as boundary conditions

$$(25)\quad \begin{aligned} r = a; &\quad H = \bar{H}, \quad M_r = \bar{M}, \\ r = 0; &\quad \text{solution is regular.} \end{aligned}$$

We introduce a new coordinate x defined by

$$(26)\quad x = a - r, \qquad d/dr = -d/dx.$$

When a thin *shell* of revolution which is not prestressed is acted upon by the loads of equation (25) we know that the stresses are essentially confined to an edge strip the width of which is of the order of magnitude $(R_\theta(a)h)^{1/2}$. It is plausible, but must be proved independently, that the same will be true for the originally flat plate, as soon as the deformations due to the loads $\bar{H}$ and $\bar{M}$ are of sufficient magnitude. This relation between the linear shell problem and the non-linear plate problem appears not to have been pointed out previously. In the equations for the edge effect derivatives of H and ϕ become large compared with H and ϕ itself and we may simplify equations (19) and (21) to

(27) $$\frac{d^2H}{dx^2} + \frac{Eh}{a^2}(1 - \cos\phi) = 0,$$

(28) $$\frac{d^2\phi}{dx^2} - \frac{1}{D} H \sin\phi = 0.$$

When $\sin\phi = \phi$, and $1 - \cos\phi = \phi^2/2$, equations (27) and (28) reduce to equations given by Friedrichs and Stoker [2].

We write in (27) and (28)

(29a, b) $$x = \lambda y, \qquad H = \mu F$$

and fix λ and μ such that the resultant equations for F and ϕ become parameter-free. Thus

(30a, b) $$\frac{\mu}{\lambda^2} = \frac{Eh}{a^2}, \qquad \frac{1}{\lambda^2} = \frac{12(1 - \nu^2)\mu}{Eh^3}$$

and

(31a, b) $$\lambda = \left(\frac{ah}{[12(1 - \nu^2)]^{1/2}}\right)^{1/2}, \qquad \mu = \frac{Eh}{[12(1 - \nu^2)]^{1/2}} \frac{h}{a}.$$

Equations (27) and (28) are now of the form

(32) $$\frac{d^2F}{dy^2} + (1 - \cos\phi) = 0,$$

(33) $$\frac{d^2\phi}{dy^2} - F \sin\phi = 0.$$

It remains to express the relevant stress and deformation quantities in terms of ϕ, F and y.

From (16), (29) and (31) follows

(34a) $$\frac{N_r}{Eh} = \frac{F\cos\phi}{[12(1 - \nu^2)]^{1/2}} \frac{h}{a},$$

(34b) $$\frac{Q}{Eh} = -\frac{F\sin\phi}{[12(1 - \nu^2)]^{1/2}} \frac{h}{a},$$

(34c) $$\frac{N_\theta}{Eh} = \frac{dF}{dy} \frac{\cos\phi}{[12(1 - \nu^2)]^{1/4}} \left(\frac{h}{a}\right)^{1/2}.$$

From (13), (9), (26) and (31) it follows, if again account is taken of the fact that derivatives of functions are large compared with the functions themselves, that

$$\text{(35a,b)} \qquad \frac{M_r}{Eh^2} = \frac{d\phi/dy}{[12(1-\nu^2)]^{3/4}}\left(\frac{h}{a}\right)^{1/2}, \qquad M_\theta = \nu M_r.$$

The boundary conditions in terms of the variables y, ϕ, and F become

$$\text{(36a)} \qquad F(0) = [12(1-\nu^2)]^{1/2}\,\frac{\bar{H}}{Eh}\,\frac{a}{h},$$

$$\text{(36b)} \qquad \phi'(0) = [12(1-\nu^2)]^{3/4}\,\frac{\bar{M}}{Eh^2}\left(\frac{a}{h}\right)^{1/2},$$

$$\text{(36c)} \qquad F(\infty),\ \phi(\infty) \quad \text{are regular.}$$

Now the terms $\bar{H}/Eh$ and $\bar{M}/Eh^2$ are of the order of magnitude of the permissible elastic strains. Fixing their magnitude we can still make $F(0)$ and $\phi'(0)$ as large as we wish by making h/a small enough. It appears from the differential equations (32) and (33) that as one or both terms $F(0)$ and $\phi'(0)$ are made large the value of $\phi(0)$ also should become large. This shows that for sufficiently small values of h/a we may, in the elastic range, have large values of the slope angle ϕ and therewith the necessity to work with the equations given in the present paper.

Bibliography

1. W. Z. Chien, *The intrinsic theory of thin shells and plates*, Quarterly of Applied Mathematics vol. 1 (1943) pp. 297–327; vol. 2 (1944) pp. 43–59, 120–135.

2. K. O. Friedrichs and J. J. Stoker, *The non-linear boundary problem of the buckled plate*, Amer. J. Math. vol. 63 (1941) pp. 839–888.

3. Th. von Kármán, *Festigkeitsprobleme im Maschinenbau*, vol. IV[4], 1910, p. 349.

4. G. Kirchhoff, *Vorlesungen über Mechanik, Dreissigste Vorlesung*, Leipzig, Teubner, 1876, 1883, and 1897.

5. E. Meissner, *Über Elastizität und Festigkeit dünner Schalen*, Vierteljahrschrift Naturforschende Gesellschaft Zürich vol. 60 (1915) pp. 23–47.

6. H. Reissner, *Spannungen in Kugelschalen (Kuppeln)*, H. Müller-Breslau Festschrift 1912, pp. 181–193.

Massachusetts Institute of Technology,
Cambridge, Mass.